The Wells Spring

A Family Story

The Wells Spring

OF SURVIVAL AND SUCCESS

Fran Sandrock

Published by Wells' Dairy, Inc.
1 Blue Bunny Drive
Le Mars, Iowa 51031

The Wells Spring - A Family Story of Survival and Success
© 2000 Wells' Dairy, Inc., U.S.A.

All rights reserved. No part of this book may be reproduced in any manner without the written consent of the publisher, except in the case of brief quotations in a review or article.

No part may be stored in, or introduced into, a retrieval system, or transmitted in any form, or by any means, (electronic, mechanical, photocopying, recording, microfilm, or other) without the prior written permission of the publisher of this book.

Inquiries may be addressed to Wells' Dairy, Inc. (above)

All rights reserved under Pan American and Universal Copyright Convention (UCC) and the Berne Convention.

FIRST EDITION

Library of Congress Catalog Card Number 95 70827
ISBN # 0-9670520-0-9

Fran Sandrock, Author
THE WELLS SPRING :
    A Family Story of Survival and Success

1. Family Memoir
2. Biography
3. Storytelling
4. Local History, Iowa, South Dakota, Illinois, Pennsylvania
5. Dairying/Agriculture
6. Social Life, Customs
7. Frontier and Pioneer Life, South Dakota
8. American Writer, Literature

> Publisher's Cataloging-in-Publication
> *(Provided by Quality Books, Inc.)*
> Sandrock, Fran.
> The Wells Spring : a family story of survival and success / Fran Sandrock. -- 1st ed.
> p. cm.
> LCCN: 95-70827
> ISBN: 0-9670520-0-9
>
> 1. Wells family. 2. Dairying--Iowa.
> 3. Frontier and pioneer life--South Dakota.
> I. Title.
> CS71.W455 2000   929'.2'0973
>   QBI99-1433

Layout by Fran Sandrock
Cover by Fran Sandrock, Ranice Skidmore
Typesetting by Ranice Skidmore, Rts Ink
Edited by Anita Singer, Fran Sandrock
Printed by Omaha Printing Company, Omaha, Nebraska, U.S.A.

The following permissions and acknowledgment pages constitute an extension of this page:

To Authors And Writers Quoted...
in this book

For the wisdom of their words, gratitude is extended. Gems that make application to today, have been expressed. Every attempt has been made to find the source, to locate copyright holders and give credit, to obtain permission for use of their words. The publisher would appreciate being notified of any omission so that proper credit can be given in any future edition(s). Grateful acknowledgment is made to:

MALTIE D. BABCOCK - (1858-1901) From *Best Loved Poems* "Be Strong," Doubleday & Co., 1936, Garden City Books 1970

P. T. BARNUM - (1810-1891) American Showman, in his eight steps to success (attributed to)

DAVID BUSCAGLIO - (1925-1998) Author of *Papa, My Father, A Celebration of Dads* (attributed to)

WILLA CATHER - (1873-1947) Nebraska Novelist of American Pioneers in *0 Pioneers,* Running Press, Philadelphia, London, Poem, "Prairie Spring"

RICHARD EDER - (1932-   ) In *Los Angeles Times* column, "End Papers" 1990

WILLIAM FAULKNER - (1897-1962) American Novelist from his Nobel Prize acceptance speech, 1959

CHIEF DAN GEORGE - (circa 1850s ?) Quote attributed to

EDWARD GIBBON - (1737-1794) Historian, in *History of the Decline and Fall of the Roman Empire,* Standard Book Co., 1957

BOOK OF JAMES - New Testament, Revised Standard Version

WILLIAM JAMES - (1842 -1910) U. S. Philosopher, Psychologist, Quote attributed to

SAMUEL JOHNSON - (1709-1784) English author, *A Dictionary of the English Language*

GARRISON KEILLOR - (1941-   ) Radio Performer, Author *Lake Wobegone Days*

HELEN KELLER - (1880-1968) Blind Author, Lecturer, in *The Story of My Life*, Grosset and Dunlap, 1905

CHARLES LAMB - (1775-1834) English Essayist, From "Valentine's Day" in *Essays of Elia* for London Magazine 1825

EDWIN MARKHAM - (1852-1940) American Poet in *Lincoln and Other Poems*, Doubleday & Page, 1912

W. SOMERSET MAUGHAN - (1874-1965) British Author, Storyteller, Playright, Essayist in *The Summing Up,* quoted in *Zen To Go* by Jon Winokur, New American Library

WILLIAM MCFEE - (1881-1966) Quote attributed to

ALFRED MERCIER - (1816-1894) American Author in *New Dictionary of Thoughts*, Standard Book Company, 1957

WILLIAM SHAKESPEARE - (1564-1616) In *As You Like It,* Act 2, Scene 1

CHARLES HADDON SPURGEON - (1834-1892) British Baptist Minister from *Treasury of Charles H. Spurgeon,* Fleming H. Revel Co. 1956

WALLACE STEGNER - (1909-1993) Iowa born writer of the American West, quote attributed to Stegner

RABINDRANATH TAGORE - (1860-1941) Bengalese Poet of India, Philosopher, Nobel Prize Winner, attributed to Tagore

STUDS TERKEL - (1941-   ) Storyteller, Radio Personality, Pulitzer Prize winner, by permission

MOTHER THERESA - (1910-1997) In Nobel Prize acceptance speech, December 8, 1979, and *Mother Theresa, Her Life and Works* by Lush Gijergil, New City Press

WALTER TOMAN - (1920-   ?   ) American Psychologist in *Family Constellation*, Springer/Verlag Publishers, New York

KURT VONNEGUT - (1922-   ) American Novelist, by permission

CHRISTOPHER'S PHOTOGRAPHY, Ltd., Le Mars, Iowa, for reproduction of Wells family photographs

THE LASTING IMAGE (Curt Strathman), Le Mars, Iowa, for reproduction of Greg Wells' family photograph

LE MARS DAILY SENTINEL, Le Mars, Iowa, for reproduction of photos

DAIRY FOODS MAGAZINE, and BILL NELLANS for reproduction of magazine cover/photograph

Published by Wells' Dairy, Inc.
Le Mars, Iowa, U.S.A.

Typesetting by Ranice Skidmore, Rts Ink

Printed by Omaha Printing Company, Omaha, Nebraska, U.S.A.

Dedicated

To the memory

of

Fred Hooker Wells, Jr. and Miriam Ralston Wells

Our Parents,

Our Grandparents,

Our Grandchildren,

Our Great Grandchildren,

and beyond

Contents

PART I

Chapters		Page
1	The Dry Days	27
2	The Encounter	33
3	The Homesteaders	43
4	The Chicago Experience	53
5	A Family on the Move	63

PART II

6	The Leave Taking	73
7	The Letter	81
8	The Beginning	87
9	Trip Back to Wellsburg	95
10	Harold, First-Born	105
11	Roy, Second Son	115
12	Harry Lee aka 'Mike'	125

13	Fun Times	135
14	Sioux City and Harry C.	143
15	The Midnight Ride	151
16	A War Intervenes	157
17	Fred D. and the Other Freds	167
18	A Frugal Life	175

PART III

19	A New Generation	187
20	Keeper of the Capital	197
21	Fire, Acid and Work Fulfillment	207
22	Wells on the Fast Track	219
23	Tough Times and Good Times	229
24	River and Spring	237

POSTSCRIPT

Long-Lived Logos	243
Ice Cream Capital (recipe)	251
Summary/Soaring	257

ABOUT THE AUTHOR

With Gratitude

The finest words in Webster's Unabridged Dictionary are inadequate to express my deep appreciation to Wells Family members for their gracious assistance and support during the time it took to bring together their memories and the fascinating story about Wells' Dairy beginnings.

For old and new photographs and pictures in this book, for maps, letters and memorabilia to make *The Wells Spring* unique and special for them and other readers, I am grateful.

To Storyteller Roy Wells, family patriarch, Keeper of Memories and family legends, for his good humor, help and stamina… a profound salute.

To Fay Wells for grace, courtesy, patience and kindness, and a thoughtful phone call that encouraged this effort, thank you.

To Fred Dale Wells for strong, quiet strength and help, my appreciation.

To Dan Wells who triggered this effort and was always there with steady support, strength and heartening words to make the journey a pleasant one, thank you.

To Gary Wells, Doug Wells, David Wells, Michael Wells

and Greg Wells, thanks for the gift of time and attention to questions, for answers and photos dug up from family and business archives to make this a better record of what came before.

To Frances Marx Wells for speaking for her husband, Harry Lee (Mike), to Shirlee Wells for photos of so many family members, especially of the deceased Harold and his family, for support... heartfelt thanks.

To Doris Zimmerman for her deep interest, contributions of old family pictures, material and information to make this book more significant, thank you.

To Clayton H. Wells and Dr. John Wells Clemens, for genealogical information and charts... accept my thanks.

To Lisa Eisma for pursuit of tasks connected with this history in a pleasing manner, full of energy and sparkle, thank you.

To Lisa Bogh, Kim Wilkens, Betty Price, Joyce Ricke, Christy Hartman and Deb Lang, I give thanks for smiling telephone faces, for earnest assistance and efficiency.

To Lois J. Prokop, Frances Fitch, Will Hampton, and South Dakota Historical Society for their interest and for historical data supplied, a thank you.

To Ranice Skidmore, more than a design artist, I owe a debt for her untiring effort, her expertise, creativity and artistry. She gave of herself to make this a special book for an exceptional family.

To Anita Singer whose editorial help and professionalism provided invaluable assistance, thank you. For her unfailing attention to editing particulars despite her own personal adversity, I am grateful.

Writing is a solitary business but I was sustained in the time it took to write *The Wells Spring* by the unusual goodness of so many.

It is the writer's privilege, as William Faulkner pointed out, "to help man endure by reminding him of the courage and honor and hope and price… compassion… and sacrifice… which have been the glory of his past."

The Wells story is a remarkable one, worthy to be set down in history, not only for family members, but for others who have a need to learn more about roots. Wells progeny right now are making history for the next generations who will want to know more about their roots and the "glory of the past."

Association with this family has been an interesting and rewarding journey that began for me in 1932 when Nellie Wells in her Beauty Shoppe on Plymouth Street SW in Le Mars, gave me my first permanent, paid for with the first money I ever earned. Screwing my hair onto rods, she doused my head with ammonia fumes, then clamped me under a monster hood. Turning on the electricity, she stood by blowing my scalp to keep me from incinerating. Thank you, Nellie.

Fran Sandrock

Foreword

*Sweet are the uses of adversity... though ugly
and venomous, wears yet a precious jewel*
                                        Shakespeare

Few Mom and Pop endeavors begin with the hard effort of one man and his wife and grow into an enormous business of international scope... in less than 100 years.

Many have gone down under the burdens of adversity, during capricious and changing economies, or the cloak of bad luck.

But Fred Wells, Jr. and Miriam, with persistence, integrity, a horse and delivery wagon, working out of the back room of their home, young children helping, were able to rise above misfortune, diminishing cash and failed tries, by placing one foot in front of the other, sometimes in anxiety and anguish, day-by-day. Day by day they built what became a dairy foods and ice cream empire.

In "Emperor of Ice Cream," American poet Wallace Stevens spoke of the death of his ideals. In his poem he held a funeral for them.

Fred Jr. and Miriam likewise were obliged as farmers and homesteaders to sack their dreams of acquiring land and a future in the hard, dry soil of South Dakota, pile their hopes for a future on a funeral pyre and move forward.

This is a book of storytelling and tales handed down by the Wells family, from their beginnings through triumph over troublesome times.

It is a record of the memories of family storytellers told as honestly as can be told about those no longer here to affirm what is told. It attempts, despite this, to maintain historical accuracy.

The Wells Spring is set down for the living sons and daughters and their progeny. Some memories will differ in the telling of the same story, but therein lies their charm. Each teller comes from a different perspective.

Many Americans, displaced by work, wanderlust or wheels, have lost their roots. No longer are we embraced by an extended family, closeby, as people were in earlier agricultural days, or when Fred Hooker Wells, Sr., and Clara, his wife, lived in Chicago before leaving all that behind for the lonely life on a South Dakota prairie.

A hunger exists among Americans for the retelling of stories from the past that can give us once again a sense of place. A nostalgia and a longing pervades to understand more, to get in touch with what shaped our lives and those of our forebears.

There is a great gift of accumulated memory in all of us, the young genius Ken Burns has said. "Hold On To Your Memories," a popular song sings away.

This book offers up the reading of both non-fiction and fiction, a weaving together of a rich tapestry of Wells tales. The purpose of fiction, one writer has said, is to tell the truth. Conversely, another has noted that the purpose, not of fiction, but of non-fiction, is to tell the truth. *The Wells Spring* author has attempted to tell the truth, to get into the heart, mind, soul and psyche of early Wells family members to bring truth and insight to the reader.

In writing we paint a word picture. In this book we also have photographical pictures. It is suggested that in reading *The Wells Spring*, family members study the pictures, concentrate on them for a period, and perhaps call up precious memories of their deceased that they didn't know they had stored away.

Mature people and nations take pride in their past, their art, literature, music, education and care of young and old. As a resurgence of interest in our past has taken place our whole country has become more adult. We begin to acknowledge what the past means to us now.

The Wells Spring takes a look into the past of those whose influence preceded the present Wells' endeavor. For this reason it may have value, also, in the education of children by providing more than looking back, but subsequently for them to view their own future possibilities.

The Wells Spring is unique not only for what Fred Wells and his lineage have accomplished. Eight family members at their industry's home base in Le Mars, Iowa, are engaged together in unusual harmony, even while entertaining differences of opinion. All the while they participate in life in a small town, toiling successfully to make an industry grow in the larger global world.

How has this happened?

Hard work, they will tell you, and clinging to principle in creating products that must not only be good, but better, the best. Pride in the product it is called.

Add the important ingredients of respect, trust, patience in working with others; the practice of virtue; humility, honesty, profiting from error, all performed in dignity… gleaned from their forebears, Fred Jr. and Miriam.

"Many things are learned in adversity," the late English clergyman Charles Spurgeon, has said. "Stars may be seen from the bottom of a deep well."

This is the wellspring, *The Wells Spring* from which an enduring ice cream castle has sprung.

Fran Sandrock

I will utter things...
we have heard and known,
things our fathers have told us.
We will not hide them from our children.
We will tell the next generation...

Psalm 78: 2-4

Part One

Fred Hooker Wells, Jr., Founder Wells' Dairy

Chapter One

Behold the farmer waits...for the precious fruits of the earth, being patient over it until it receives the early and the late rain...

James 5: 7-10

The Fred H. Wells, Sr. family lived in a home that housed a general store, the Wellsburg, South Dakota post office and dance hall. The structure also was used as a church. Farm buildings on the left served Wells, the farmer. Fred Sr's pay as postmaster for the year 1909 was $152.00. In the drought year 1911 it was $120.00

(Photo courtesy of Doris Wells Zimmerman)

The Dry Days

Fred Hooker Wells, Jr. looked up at a pale hot sky showering heat on the land below. Then he looked down at his dry, dusty acres. Wells was troubled.

Dust devils swirled in daily on the back of strong hot winds from the southwest. Dust covered furrows between corn rows, clutched the feet of each cornstalk that struggled all the long summer to reach for air and water.

Every morning the sun glared at young Fred Wells and his family across the open prairie. By evening the red sun grown redder trailed over the horizon like a meteor burning itself into their memory. Next day the wind blew hotter than before. The air grew drier, dustier.

Rain had not fallen in Wellsburg, South Dakota in that Summer of 1911. Roy Wells, Fred's son, at 89, remembered hearing the oft-told story. He was two at the time.

"Corn that year was two, three inches high in October. It didn't rain much in South Dakota… any year. No rain that year."

Nothing. Not a drop. Clouds blew in occasionally promising an end to the drought. Looking for the end to it was endless.

Short cornstalks had lost their green color. Their thin, limp leaves fluttered and crackled in the wind.

Fred Wells' field corn had not produced enough through that parched earth to feed his livestock. Cattle bellered in the cattle yard for feed and for water that had to be hauled by wagon twenty five miles from Philip to Wells' water tanks. Thirsty cattle trampled each other to reach filled tanks. The sound of those cattle disturbed his sleep.

Six years ago Fred had come here from Chicago to homestead with his parents. With them came eight other children Harry C., Howard, Raymond, Jessie, the oldest girl, Florence, Stella, Ella Mae and Nellie the youngest. Fred Jr. at 20, had taken Miriam to wife not long before they left Chicago.

Fred turned to look toward the home his parents had built. He had helped to build that home. The part sod, part wood frame dwelling housed ten members of the Fred Wells, Sr. family. (Fred Jr. and Miriam, who lived nearby, had had two sons since coming to Wellsburg; Harold, their first born, and Roy, their second son.)

Fred looked at the church his father had built. With his son's help. Attached to the house, a general store and post office, and a dance hall, had all been constructed with the family's help. At the dance hall they were refreshed after many a difficult day… with music, companionship and laughter.

I am part of all this, Fred thought, I have had a hand in creating this. It hasn't been easy, but there it stands. Proof that you can homestead, you can make a home for a family out here on this treeless prairie.

Miriam and I put everything we had into plows, horses and wagons to start farming in this new country, where the prairie sod is hard as rock. But this land, three sections, doesn't produce enough for two families. Not without rain and crops. Barely enough to feed and clothe us. Those growing boys need shoes. Winter is coming.

To scrape together enough money to buy seed corn to seed this land next spring will be hard. And if we do seed it, will the rains come?

His eyes scanned the southwest once again looking for a sign. Just a small sign. He stood like this daily in his field, looking up, praying for rain to fall. But rain hadn't come. God, rain hadn't come.

What were they to do? Would they starve out here in this "god-forsaken" country as Miriam called it? If God were here there would be trees, wouldn't there?

The bad land nearby looked as if the devil himself had come through with a rake and a stick to beat it up. Beauty? Where is the beauty? Only heat and dust and bitter winters where the wind never stops blowing.

Dust blew into their home to be swept out with a broom. Dust settled in the bed clothing at night to be shaken out each morning. Dust blew into wet clothing hanging on the clothesline. Monday morning's wash, done by hand. Where was the beauty? Not in their coffee cups to be washed of dust before morning coffee could be poured.

Christmas is coming. What kind of a Christmas will we have? The father of two little boys and a good wife should be able to provide for his family. But there's an account at the general store for groceries owed.

    Miriam had always supported him in what he planned, what he wanted to do. He hoped his heart would not break when he spoke to her. He must be strong.

    "Miriam," he would say. "I think we better get outa here."

(Courtesy of South Dakota Historical Society)

Chapter Two

Not many sounds in life exceed in interest a knock at the door

Charles Lamb

Part sod home of Fred H. Jr. and Miriam, in South Dakota 1908. Fred Jr. holds baby Harold. Women are said to have loathed the "soddies" since the roofs leaked, dirt, bugs and mice fell through. Soddies were heated with buffalo chips

(Photo courtesy of Madonna Wells Searing)

The Encounter

Miriam Wells was startled by a pounding on the door, an insistent pounding, that sent through her a shot of terror. She was alone except for her two small boys. Fred Jr. had gone to the "big river, (the Cheyenne,) with his wagon to pick up wood for the winter," Roy, a toddler then, recalls the story.

The blowing wind carried voices of an approaching crowd. Miriam peered carefully through her curtained window.

Coming down the dirt road, ahead of a heavy cloud of dust, a large number of Sioux Indians, warriors first, some on whinnying ponies, others walking, camp dogs pulling possessions, squaws and crying babies, drew near.

The pounding grew louder. Should she open the door? Would they force their way in if she didn't? What did they want?

Were they going to take her boys? She picked up Roy, grasped Harold's hand and pushed them behind the big black cook stove. Grabbing a blanket, she covered them.

"Don't make a sound. Hush," she whispered. "Not a

sound. Be quiet."

Neighbors said the Sioux, the Cherokees, the Comanches have stolen young children. Would they steal Harold and Roy?

The white man has killed off their buffalo, taken their land. How will they treat us? How will they treat me?

Dear Lord, don't let them take my little boys.

Should she hide? Where? If Fred were only here. He had taken the rifle with him. He might see rabbits or squirrels down by the river. What would she do if these strangers broke in?

She opened the door. Just enough to answer the pounding. A great gust of wind wrestled to take the door out of her grasp.

An Indian chief stood facing her. Tall, erect, he wore a headdress of feathers that poured down his back and trailed the dusty ground. His buckskins were fringed, his face painted.

The blunt end of a feathered tomahawk was pointed at her and the door. He was ready to pound again. He looked at Miriam. His brown fingers pushed at his open mouth while he made a chewing motion.

What does he mean? What is he saying? They must be hungry. Their corn has burned out this year. Just like ours.... Buffalo are hard to find. Winter meat is scarce. But what can I give them?

I must find something. I can't tell him we have no food. They might take the boys instead. Oh, Lord, don't let them take our boys.

"She was scared to death," Roy remembered 87 years later.

His mother had heard stories of Sioux Indians on the move. It's fall and they are moving south, she decided. What shall I give them?

Lowered into the well, a pudding cooled. Nodding to the chief, she pointed toward the well and made the same chewing motion. He moved to the well.

She and Fred had bought squash, potatoes and dried beans, that came from Philip to the Wellsburg general store, run by her in-laws. There were a few prairie turnips. And today's homemade bread. No eggs were left after she finished baking. In the food cupboard dried meat lay carefully wrapped.

"She gave them everything she had," Roy says.

When at last they left Miriam closed the door and secured the hasp. Shaking, she sat down on an old rocker and cried. She felt weak and helpless.

But the children mustn't see that she was frightened. They would be listening. Wiping her eyes on her apron, looking out the window for assurance that the vistors had gone she turned to the stove. Behind it two boys huddled in terror.

Yanking off the blanket, she tugged at the boys and hugged them hard. Oh, the loneliness of this tree-starved country. If it weren't for the boys and her in-laws a half mile away, how could she stand it?

She was happy with Fred Jr. But most of the time he was out in the field plowing, planting, raking, hoeing, mowing, cutting wood, or in the barn tending the animals.

Until she met Fred, Miriam Ralston's Chicago life had been lonely. She missed her mother.

In 1896 when she was nine her mother died. Her older sister, Maggie, 30, and her brothers Archie, Willie, and Charles looked after her. Her brothers and her sister were born before the Ralston's emigrated from Scotland. Miriam was born in America. Their father, a drayman, owner of a business on Lake Street worked hard. Just like Fred Jr.

Several years after the death of her mother, Father Ralston suffered an accident. His dray horses, frightened by a noisy elevated train, panicked throwing Ralston into a steel piller. He died, leaving Miriam orphaned.

Oh the loneliness out here, Miriam thought, as she rocked the boys on her lap and listened to the wind outside. I hate that wind, the taste of dust, the smell of it. I can't stand it here. I'm afraid.

I don't want Fred to know…how much I want to go back to Chicago. I can't tell him . But how can I keep it from him?

What would he think? His parents have been good to me. Fred won't want to leave them. What would they think? About me? His father depends on my husband. Especially since Harry C. went to Doland… homesteading. That would leave Ray and Howard to help Fred Sr. And the girls. They are all big and strong.

Our money… is running out. Maybe there won't be enough for seed corn and spring planting. Fred needs a coat. The boys need winter coats.

Oh, God, I want to leave this place.

What would he say if I said to him, "Fred, I think we ought to go back to Chicago."

Miriam Ralston at her eighth grade graduation in Chicago

(Photo courtesy of Wells Family)

The Encounter

Mr. & Mrs. Gustav Lobel
request your presence at the marriage of
their sister
Miriam Ralston
to
Fred A. Wells, Jr.
Wednesday evening, the sixth of February
nineteen hundred and seven
at eight o'clock
at their home, 2532 Superior street
Chicago, Illinois

Fred Jr. and Miriam Ralston were married in Chicago February 6, 1907 before they moved west that year with Fred Sr. and his family to South Dakota

(Photo courtesy of Dan Wells)

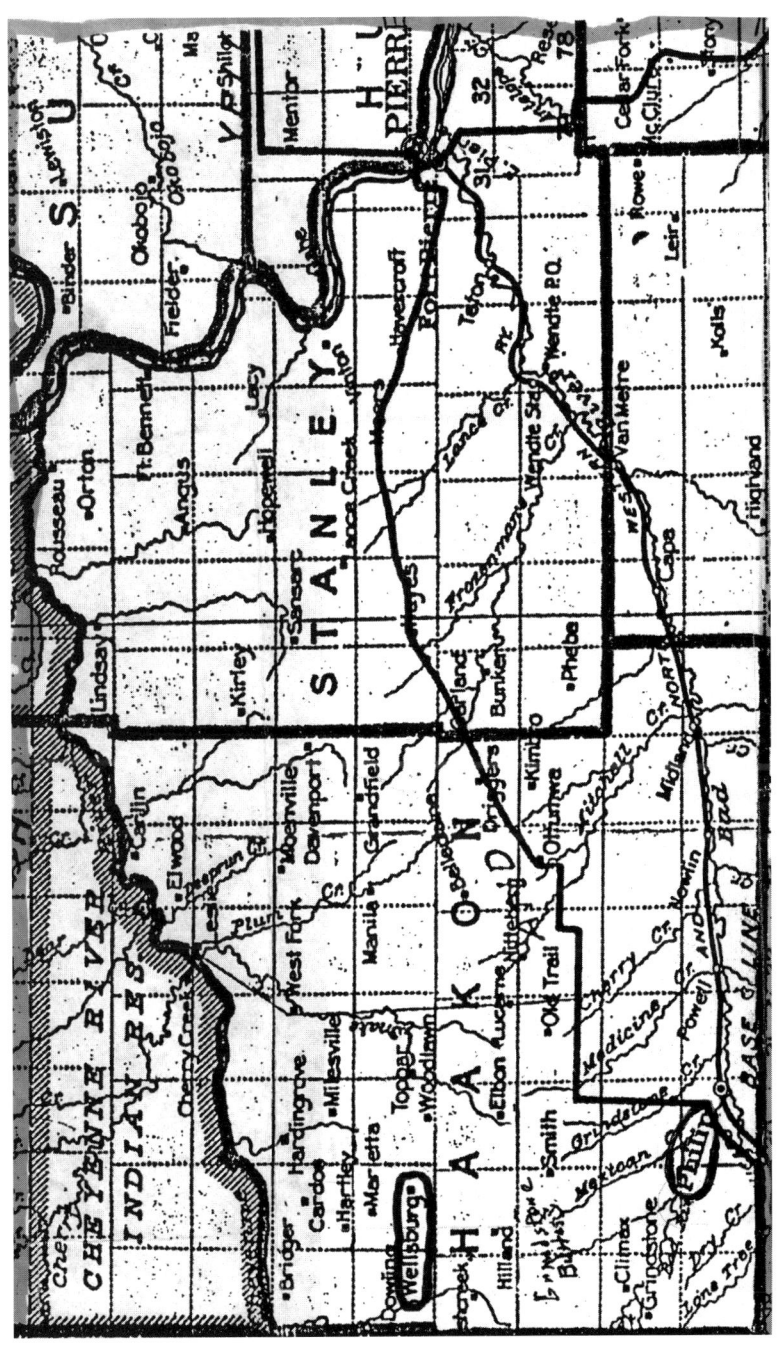

The Cheyenne River, where Fred Jr. had gone for wood, lay north of Wellsburg. South through the Grindstone Buttes about 25 miles lay Philip where supplies were purchased and brought back in a wagon pulled by four horses. Pierre, 85 miles east to the closest railroad, was considered to be on the eastern edge of the Great American Desert

(Photo courtesy of Lois J. Prokof)

Chapter Three

*...every good that is worth possessing must be paid for...
in strokes of daily effort...*

William James

Fred Sr. and Clara Marvin Wells and their family of nine children are shown when Nellie was a baby (held by her mother) in 1896. Fred Jr. is top left, and Harry Cole Wells, top right

(Photo courtesy of Wells family)

The Homesteaders

Fred H. Wells, Sr., and Clara, his wife, rose from their hard, backless wagon seat as they left Pierre for their homestead.

"Look! Look!" they shouted to their children, nine of them seated on hay in the high-sided wagon bed, "Here it is! Our home! South Dakota!"

Ahead lay miles of tall buffalo grass hip-high on a prairie, sometimes gently rolling, sometimes flat as a stove iron. Thirty miles to the north lay the Vermilion plateau. A few farmsteads sparkled in the distance through the sun-warmed clear cold air of Spring, 1907.

Tales of the old Deadwood Trail ahead and the river bluffs once "black with buffalo, turkey, deer, elk, coyotes, wolves"… and Indians that peppered these grasslands now deeply rutted with roads…had fired their fancy.

Stories about wild strawberries, red and black chokecherries, buffalo berries, magpies, eagles… had had a powerful pull. Grapes, plums, wild pears, blackberries, black walnuts along the river, what a prospect!

"We're here, do you hear? We're here!"

On this glad day nine children and Fred Jr.'s, wife, Miriam, (whom he had married in Chicago February 6,) all rose and shouted, laughing with the delight of it.

The hard trip from Chicago was almost over… sleeping now in the wagon, men on the bare ground, a canvas cover protecting the wagon. The long train trip by Chicago and Northwestern rail to Pierre was behind them.

"Whoa! Whoa!" Squinting at this new land, Wells pulled his team to a halt. Shading their eyes, twelve travelers looked ahead as far as they could see. This was to be their home.

Hadn't Teddy Roosevelt called the West a land of opportunity and adventure? But time was of the essence. Soon there would be no land left to be claimed. Hadn't they seen wagon after wagon along the way west, pots and pans hanging from sideboards, gunny sacks and bags loaded with possessions?

"Giddap, giddap, we're here!" Fred Wells clapped leather reins on the horses' backs. The large, over-sized back wagon wheels crunched. Belongings hanging on the sideboards clattered.

They still must ride to find land not already claimed.

Here was sunshine every day of the year, and a chance to build a home, even a town. And to acquire land, a farm for each child.

Harry C. had turned 22, Fred Jr., was 20, Ray would be 18 soon. Howard had just had a fifteenth birthday. The girls ranged in age from Jessie turning 19 to Nellie, 9. Able-bodied sons and daughters could handle the work.

In Chicago Fred and Clara had decided to make a change. He would be 45, she, 49. Chicago had been good to them in some ways, but South Dakota would be better. The two sold their dairy business and some of their possessions. They had talked to neighbors and friends.

A man needed about $1,000 to homestead and get a good start. The time was right. They could take the Chicago and Northwestern Railroad to Pierre. The train would be better than traveling by horse and wagon, although the children must sleep on boards across the train seats. At Pierre they would buy horses and a wagon to make the trip to their homestead.

Arriving at Pierre in South Dakota east of the Badlands, in late spring of 1907, they still had eighty five or more miles, southwest, to cover. Once a suitable spot was found, they could stake their claim.

To provide shelter at their claim site was their first task. At Philip they saw the new John Deere breaking plow. Better to buy it than to hire a sod breaker at up to $3.00 an acre.

Eagerly they hand-dug a hole to help get the plow into the hard ground. A dugout in which to live came first. Strips of rock-hard prairie sod and grass, two to four feet deep, were pulled out of the ground and piled like a brick wall. A large one-room home of prairie sod roofed with tree limbs and straw emerged. Thick walls would keep the home cool in summer and warm in winter. Trunks, bags and packing cases served as furniture. Work went fast with so many helping hands.

In yard-long steps, the men paced off their claim: eighty four paces to an acre seven hundred fifty paces for 160 acres. Each man over age 21 was eligible to claim acreage. Marking each corner of the claim with rock, stake, or mound of dirt gratified the soul with a feeling of possession.

Fast work was essential. The growing season was a short one from May 10 to September 15, The first killing frost could rob a man of a year's toil.

A sod shelter for cattle, and water tanks were needed, as well as a well-driller and a windmill, if longed-for water was to be found.

Into the broken soil sod corn was thrown every few rows. Sod corn provided up to 30 bushels an acre from new soil. Turnip seed, squash, bean seeds were scattered and potatoes planted. To lay in provisions that would take them through the winter became the next goal.

The elder Fred Wells' dream to build a town began with the frame home, attached general store and post office. The store was used also as a church and dance hall. This he called Wellsburg, South Dakota.

Neighbors were miles away, the nearest railroad 85 miles, Nellie Wells wrote later. The family made friends in this friendly country. On weekend evenings fiddlers played old tunes to laughter and dancing: "Quilting Party,' "Wait For the Wagon," "Skip to My Lou, My Darling:"

> *Flies in the sugar bowl, shoo, fly shoo…*
> *We'll keep this up until half past two…*
> *Can't find a fat gal, skinny one'll do…*
> *Skunk's in the parlor, phew, phew, phew …*
> *Flies in the sugar bowl…*
> *…two by two…*

Clara and her daughters baked thirty cakes for midnight lunch after dances every other Friday night. Cost 25 cents a cake. Plays and programs were held on alternate Fridays. Nellie Wells Chenhall and Stella Wells Hampton recounted this in a book *Prairie Progress in West Central South Dakota*.

When harvest time came in succeeding summers it did not yield what the growing family needed. Fred Jr., and Miriam had produced two more mouths to feed, Harold's and Roy's. Harry Cole had gone to claim better land at Doland, South Dakota.

The rich new soil gulped up moisture. Scanty rainfall had not reached the expected 20 inches each year.

The land, they learned… the land they had chosen… was semi-arid.

In 1923 Fred Sr. sold his holdings, except for the land, and moved from Wellsburg back to Le Mars, Iowa. He and Clara had gone to Le Mars with Fred Jr. and Miriam in 1911 but they returned to South Dakota five years later. Here a farm sale is being held. That's Fred Sr. in front of the screen door. Ella Mae Wells and her two children and Stella Wells and Will Hampton stand to the right of Ella. Note that a basement had been added to the home, now sided. Cellar door is at extreme lower left

(Photo courtesy of Doris Wells Zimmerman)

Fred H. Wells, Sr. and Clara Marvin at time of their marriage

(Photo courtesy of Wells family)

The Wells Spring

*Shown is the lower right hand portion of
5-19 Homestead Township Map, Township: 5N, Range: 19E
Stanley County*

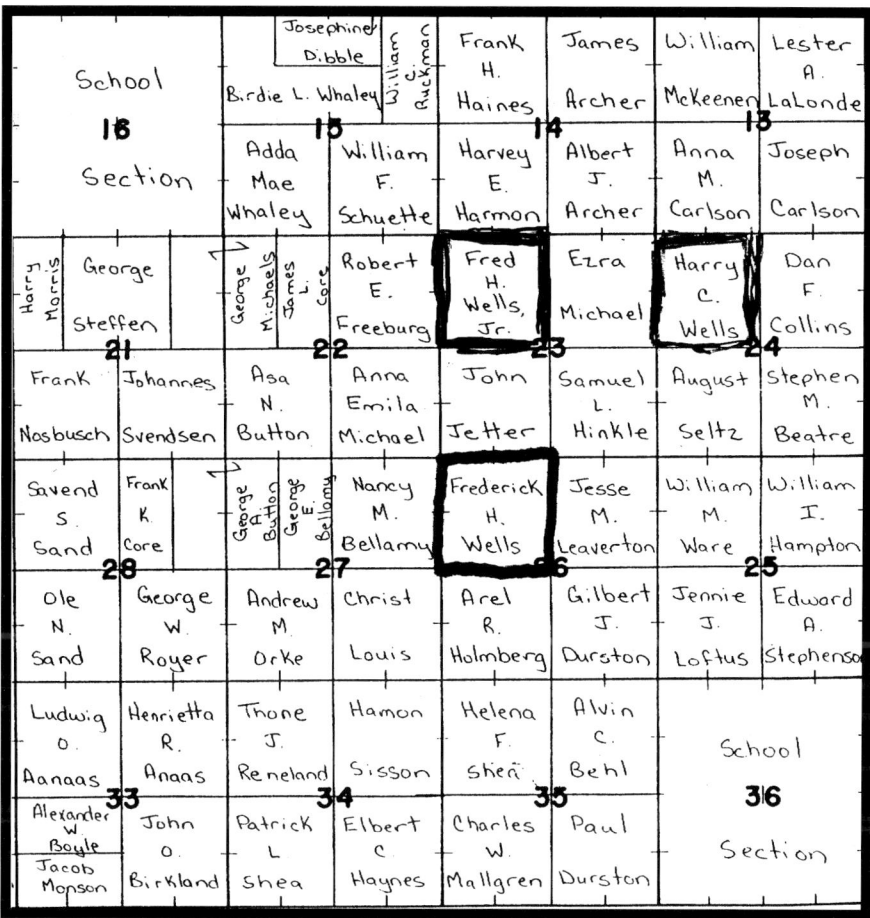

Range map shows where Fred H. Wells, Sr and his sons, Harry C. and Fred H. Jr. homesteaded. Fred Sr. and his family built a sod schoolhouse where children stayed in several shacks during the week, then went home for the weekend. Carrie McKean taught 35 students, grades one through eight

(Courtesy of Lois J. Prokop)

Chapter Four

...a person's family represents the most influential context of his life... It exerts its influence more regularly, more exclusively and earlier in a person's life... than do any other life contexts

W. Toman

In a 1903 photo Fred Jr. is shown ready to deliver milk for the Fred Sr. business on Park Avenue in Chicago. The business was sold to Cunningham Ice Cream Co. before the Wells family moved to South Dakota

(Photo courtesy of Wells' Dairy, Inc.)

The Chicago Experience

 Grandfather Fred Hooker Wells, Sr., arrived in Chicago in 1877 when he was 16. Nellie Wells, his youngest daughter, told that story in a letter before her death. (See her letter at the end of this chapter).

The Wells Family has been on the move since the 1600s. Storytellers in the family have handed down the Wells tradition of movement. How they got to Chicago is a story that Roy Wells, the family's storyteller and collector of memories, recounts:

"They came in a bunch from England. One bunch went into the foundry business in Connecticut… moulder of small items… salt and pepper shakers… others went into the dairy business.

"They moved to Vermont and to Western Pennsylvania. In groups. Then to Chicago."

Roy heard this as a young person from his elders.

Fred Sr.'s young life had been full of movement. Born of Shepard and Caroline Silsby Wells, November 22, 1861 in

Springsboro, Crawford County, Pennsylvania, one of four children, he was orphaned at age seven.

His father, Shepard, died in 1866. Shepard first married Sarah who left him (in death) with six children. Caroline Silsby, his second wife, died (in 1868,) two years after her husband.

Grandmother Silsby took Fred Sr. in for two years after Caroline's death, his Uncle Obed Wells in Conway, Massachusetts cared for him two years; his sister Lidie for a year in Springsboro, Pennsylvania; then his Uncle John Cole with whom he moved from Keepsville, Pennsylvania to Chicago. Nellie Wells told and retold that story.

By 1881 Fred Sr. had met and married Clara Marvin, a Chicago girl. Born in Bridgeport, Connecticut, she lived in Canon, Connecticut and moved with her family to Chicago. Here she met Fred Sr., her future husband. They married July 3, 1861.

"At Oak Park, Fred Sr. went into the milk business," When? Roy doesn't remember hearing. The date is unknown.

"Milk was shipped into Chicago from other dairies… everything by rail," coming in ten-gallon cans from the Wisconsin and Illinois countryside, Roy tells.

"They delivered door-to-door in Chicago, with a horse and wagon. Other dairies just brought milk in and sold it. Fred Sr. did his own bottling," establishing a practice of doing the job better than competitors, a trait handed down, like family stories, to Wells offspring.

At 45, Fred Sr. was not satisfied with Chicago. A restlessness to move ahead, an itch born of his ancestors, gnawed at him.

The Wells Spring

The Chicago experience had been troubling. The Great Chicago Fire of October, 1871 was still on the minds of Chicagoans when Fred arrived with his uncle. Burned into their memory, that fire killed 300 and left 90,000 homeless. Other smaller fires took more of the city's frame buildings, after the big fire.

A national railroad strike involved Chicago rail lines; the Haymarket Riot, a labor dispute of 1886, saw bombings in which five policemen were killed and four labor leaders hanged. A cholera epidemic raged. All this made Fred and Clara uneasy.

Into this time of unrest, Harry Cole Wells, their first child (one of nine) was born in 1883. Fred Hooker, Jr., followed in 1885. Their first daughter, Jessie, came in 1886 and Ray Shepard arrived in 1887. A fourth son, Howard Ira, was born in 1890.

Ella Mae made her appearance in 1891, Florence Mary in 1892, Stella Carolyn in 1895 and their last child, Nellie Lydia, in 1896.

During this time an Iroquois Theatre fire killed 571 in Chicago. Another railroad strike brought federal troops to the city to quell rioters.

Better pay for laborers was a prime objective for agitated strikers. Farmers, too, were unhappy over milk prices and railroad rates. By the late 1890s one million people lived in Chicago, creating a big, busy, crowded city.

President Theodore Roosevelt after visiting the Badlands of North and South Dakota lauded that country. Here, under the Homestead Act of 1862, title and possession was promised the homesteader. An enthusiastic, ebullient President persuaded many a pioneer to go West. He promised fulfillment of dreams. Public land could be acquired within five years after living on the land in productive occupancy and paying a small fee: a powerful stimulant for homesteaders.

 Many from Illinois were moving West for a better life.

 Land could be theirs, Fred Sr. speculated, theirs by following government directions. Couldn't it?

 But their children were in school, some were working, Clara said, earning money. A mother thinks of these things. Doesn't she?

 The sons and daughters could help on a South Dakota farm, couldn't they?

 The sons are learning the milk business here in Chicago, Clara reasoned, aren't they?

 The sons, and daughters, too, when they marry, could acquire land in South Dakota, Fred Sr. argued.

 But there are phones here, and hospitals, and doctors, aren't there? What is there out on the prairie?

 Isn't the prairie a better life than this busy, disaster-filled city?

 But think of the children? What of the children?
A mother worries about these things.

Undated Letter From Nellie Wells

Ma was born on May 25- year 1858 in Bridge Port Conn. she lived there untill she was 18 years old then moved to Canan Conn where she lived 2 years and joined the Epistaple church. then moved to Chicago where she met her husband Fred H. Wells and was married to him July 3rd 1881. They lived in Chicago about 33 years, and raised nine children. They moved west to South Dak. in the spring of 1907 on filed on a Home stead, which was a very hard life as it was 85 miles from rail road. And in the fall of 1911. They moved to Iowa on a Farm 8 miles south of LeMars Iowa for 9 years, they moved back out to S.D. again and lived there untill the spring of 1923 then they had a sale on March 30th 1923 and visited with Children in Doland S.D. LeMars & Chicago, where Nellie took up beauty work and moved to LeMars and have been in business here for 7 years.

3

Pa was born on Nov 22 - 1861. Your father Shepard Wells died when you were 2 years old. Your Mother died when you was 4 years old and her name was Caroline Silsby. Then you went to your Grandmother Silsby and lived with her about 2 years then with Brother Obed about 2 years then with sister Lidia about a year she lived in town Springboro Penn. Then from there you lived with Uncle John Cole and till about 16 years old, you went to the Meleth church at Keepville Pa. while living with Uncle John Cole. Then he went to Ohio. when he was about 16 years old and was married to Mother, Miss Clara Marvin in July 3rd 1881. 9 children were born to this marrage while living in Chicago. Then in the spring of 1902 moved to S.D.

(Letter courtesy of Doris Wells Zimmerman)

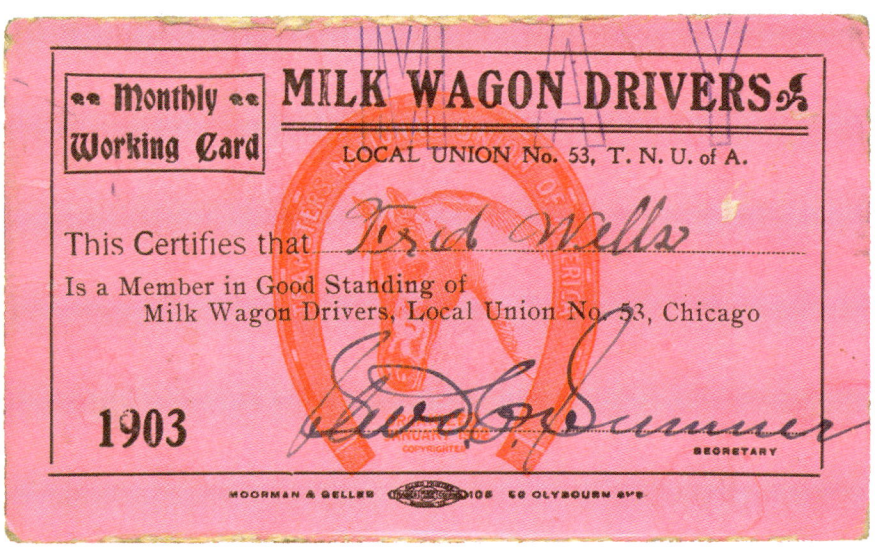

Chapter Five

I look... to the history of the family as giving a person a sense of place, that we were not just chips floating on the waves, that in some way we were meant to be here, and had a history. That we had standing

Garrison Keillor

Welles Will

In his will, dated 9 February 1637, Thomas Welles referred to his wife, but not by name, and to his three minor children, Thomas, Mary, and John. To his eldest son, Thomas, he bequeathed his house, "after the death of his mother." This confirmed suspicion that the house mentioned in the will of Thomas Coleman (Frances Wells' second husband) had been the rightful property of Frances and her son in the first place.

The will (Consistory Court Of Worcester, 1637, No. 145, on FHL Microfilm 98,054) is abstracted as follows:

> In the name of god amen be it knowe to all men that I Thomas Welles of Evesham weaver doe make my last will and testament the ninth day of feberrary 1637 the maner and forme as foloweth
>
> first I beequeth my soule unto the hands of my lord and saviour Jesus Christ whoo hath Redemed it next I bequeth my body to the earth and all my worldly goods In maner and forme following
>
> first I forgive my father the 5 pownd which he oweth to me and I give to my father 6 pownd more to be paid in three yere by equell somes fourty shillings a yeare but if he dy the mony that is unpaid to remain to the Exseckiter.
>
> Next I give to my eldest son Thomas 20 pownd to be paid at the age of 21 yeares and my house after the death of his mother Item I give to my daughter Mary 30 pownd likwise to my son John 30 more but if my wife be with child and it live then it is my will that 10 pownd a pece shall be taken from John and Mary and given to it and it is my will that my son John shall be paid at the age of 21 yeare and my daughter Mary at hur day of mariage or at the age of 21 yeares: but if they prove stouborne and dissobedent then it shall be left to the will of thir mother and the overseers when they shall have it: and further it is my will that if my son Thomas dy without a aire then it is my will that it shall come to my son John and if John dy without a aire then to com to the other son if it be a son or elce to remaine to the Daughter if ther are two or elce to remaine to my daughter Mary. Item I give to my man Charles Whitell a shipe & hoge worth eight shillings or two hachibs which he nowe doth work with upon his good behaviour to his dame.
>
> Item I give to my godsons Joseph Blissord and John Welles 2 shillings a pece.
>
> Item I give to John Pathit 2 shillings. Item I give to Ann Albright and Jone the daughters of John Allbright 2 shillings a peece.
>
> Item I give to John Allbright and Catherin the sonne and daughter of Richard Allbright 2 shillings a peece.
>
> Item I give to John Loe and Sara Loe the son and daughter of Griffen Loe 2 shillings apeece.
>
> Item I give to Sara Ordway 2 shillings.

(Courtesy of Clayton H. Wells)

A Family On The Move

The Wells succession for 400 years has been the story of a people on the move… adventuresome, risk-takers, many of the men long-lived with courage to explore opportunity in a new land, charge into new ventures, plow untried territory. To face challenges in a changing time has been their legacy.

Clayton H. Wells, a genealogist in Florida, has traced the Wells seed back to the early 1600s. Take a look at the family history:

Thomas Welles born in Evesham, Worcester County, England, a weaver, "possibly of hats," died in 1637. According to early records, Welles executed a will February 6, 1637 and may have left property, enough to be dispersed. His birthday, however, is not recorded. (See opposite page)

To America

Thomas Wells II (now Wells) emigrated to America about 1640 with his mother, Frances, widow of Thomas Wells I, and three other children.

A Family on the Move

Young Thomas, a teamster was killed in the "Battle of Muddybrook" September 18, 1675 in Hadley, Hampshire County, Massachusetts. He was 47.

Thomas II and his wife Mary, an Englishwoman born in Derbyshire, produced 15 children before his death.

It is from this generation that the Fred Hooker Wells, Sr. family from Chicago and **Wellsburg, South Dakota** is descended.

Noah Wells, child of Thomas and Mary, begat and born in Hadley, July 26, 1666, moved with his family from Hadley to Deerfield, to Lyme then to Colchester, Connecticut, where he died in 1712, at age 46. Noah left a wife Mary and seven children. Of Noah's seven, one was Samuel.

Samuel and his wife, Ruth, had a family of five. Samuel was on the roster of Fort Massachusetts 1746, suggesting that Samuel was associated with the military. He was born in New London County, Connecticut in 1704, baptized in Colchester and died in Colchester at age 42.

Samuel II their son, was born while they lived in Massachusetts.

Wells progeny showed a bent for the spiritual as well as the military. Samuel Wells II became a church deacon and selectman in Conway, Franklin County, Massachusetts.

Samuel had first married a third cousin, Bathsheba Wells. After one child and her early death, Samuel married Hannah Weaver. There followed nine children.

By the time of his death at age 75, June 6, 1804, the family had connections in Vermont. Samuel II is buried on **Wells Island** (in the Lamoile River) at Georgia, Franklin

County, Vermont. Wells family members often named communities after the clan.

To Pennsylvania

Samuel III, a child of Samuel II and Hannah, was born at Hatfield, Franklin County, Massachusetts.

A farmer and salt miner, Samuel moved from Massachusetts to Crawford County, Spring Township, Pennsylvania. He married Sylvia Allis in Deerfield, Massachusetts. The two brought forth ten children.

Erie County, Pennsylvania records show that Samuel and Sylvia, who died two years apart, were first buried on their farm before interment at Hope Cemetery in **Wellsburg, Pennsylvania.**

Obed, one of Samuel and Sylvia's ten, fathered fourteen children. Through his first marriage to Phoebe Thomas that ended in her early death he had ten children. Obed married Dorcus, Phoebe's sister, after Phoebe's death. Dorcus bore four youngsters.

(Women of those early times tended to die in their thirties and forties.)

Shepard B, offspring of Obed's marriage to Phoebe was the father of **Fred Hooker Wells, Sr.**

Shepard, born in 1820, a Pennsylvania farmer, first married Sarah Silverthorn who bore him six children. Sarah died at age 35.

Shepard then married Caroline M. Silsby who lived 43 years. Born to them were four children.

Shepard himself lived to only 46 years. He is one of the

few early Wells descendants who remained all his life in the place of his birth.

To Chicago

Fred Hooker Wells came to this life November 22, 1861 on a farm in Springsboro, Crawford County, Pennsylvania.

After venturing to Chicago as a dairyman he married Clara Marvin July 3, 1881. Clara Marvin was born into a family of Episcopalian Church members in Bridgeport, Connecticut. To her marriage with Fred were born nine children, all in Chicago.

Together they founded **Wellsburg, South Dakota.** In Dakota, Fred Sr. became a postmaster/merchant/farmer, a fifth generation Wells to cultivate the earth. Wellsburg no longer exists, having been torn down about twenty years ago when the property sold.

At retirement the couple moved to Le Mars, Iowa. Both are buried at Le Mars Cemetery in a family plot. She died at 74, he at 76.

Of Fred and Clara's offspring… those descending from their son **Fred Hooker Wells, Jr.**… more have remained in one spot than almost any other descendants in the family.

The moves of the rising generation have been vertical, as well as across states, countries and boundaries. Their achievements have been phenomenal.

Shepard Wells and Carolyn Silsby Wells

(Photo courtesy of Wells family)

Part Two

Chapter Six

The important thing is this: to be able to sacrifice at any moment what we are for what we could become

Anonymous

Most transportation in 1911 was by team and wagon. Miriam and Fred Jr. made the long trip from Wellsburg that fall by wagon, with Fred Sr. and Clara (Photo courtesy of Wells Family)

The Leave Taking

A cold November wind howled and whined in melancholy. A storm door, torn by the wind's force, flapped crazily. Wind banged a window behind them.

It had been a hard, dry South Dakota summer and fall. Dust whirled in little circles out in the fields. Horses hitched to a wagon hung their heads, their tails whipped between their braced hind legs.

"Are you ready, Miriam?" Fred Jr. called to his wife trying to be heard over the wailing wind.

Down the road a half mile, where Fred Sr., and Clara lived with the rest of the Wells family, came the two grandparents clutching their suit cases. They, too, were leaving with Fred and Miriam. Jessica, Florence, Stella, Ella Mae and Nellie followed, hugging themselves, aprons and jackets wind-wrapped in circles around their bodies.

Gusts of wind carried their "Goodbye! goodbye!" ahead of them, down the road east.

Jessica had married Philip Kenaley three years before. She came, nevertheless, for the leave taking. Howard and Ray, her brothers, already at work, had said their goodbyes last night.

"Will you be warm enough?" Clara inspected the straw-packed wagonbed where Harold and Roy, encased in flannel and wool, huddled under heavy blankets.

"Yes, I'm ready," Miriam came out carrying more old quilts and horse blankets.

"Write to us," Nellie said.

"Let us know when you get to Chicago." Jessica leaned over the wagon's high sideboards to reach and pat little Roy and Harold. Jessica's last chance, Miriam thought as she tucked her boys in tighter and settled down beside them.

She was not feeling well, but she must not let Fred Jr. know. After all, they had talked about this many nights before deciding it was the best thing to do: go back to Chicago and the dairy business, where she had been a teacher.

Miriam felt a little sad. She wasn't sure why she was not happy about leaving their farm. Maybe she was in the family way again. She hoped not. She hoped Clara would not notice on their long journey together.

She was glad not to have another summer and winter in dry country. She was glad to leave the wind and the dust devils and the "wind sickness,"… to leave this land of wolves.

To leave this land of wolves and Indians and snakes! Why, her mother-in-law not long ago had opened her door and nearly stepped on a snake. A large snake, it must have been a rattler, Clara told her.

Curled up in the sun, it looked up at her with cold, hateful eyes, its tongue lashing out at her, in and out, in and out. Clara stood there paralyzed for one horrible second, then slammed the door so hard that Nellie came running.

Before Clara shut the door, she told Miriam, she got a good look at the snake. Big, with a round thick body and long tail, its scales sparkled in the sun.

"Did you hear it rattle?" the men asked.

"No, I didn't wait long enough." They laughed at that. Probably a black snake or a bull snake, they said.

But Clara wasn't so sure. She had heard about rattlesnakes on the plains and feared going out her front door afterward.

It felt good to cross the Big Sioux River into Iowa at Akron, and head eastward. Fred Jr. and Miriam, Fred Sr. and Clara, had spent six years… they needed only five to homestead and claim the land… in South Dakota. Now in 1911, at least they were not giving up what they had worked so hard to get as their own.

Howard and Ray could carry on. The four girls still at home would help.

The travelers left with what they believed was enough money to reach Chicago. Soon they were low on funds.

Travel fatigued them. The raw weather was not abating. The little boys were restless. "Are we there yet?" they kept asking.

In Plymouth County, the four adult travelers noticed a change in the land that pleased them. The soil looked better than the hard, dry sod they left behind.

In Le Mars, they saw a bustling downtown district and

fine old Victorian homes. The place appeared prosperous. Fred Jr. smiled. This made Miriam smile, too.

"I wonder," said Fred. "I wonder if we couldn't stay here for a while. Someone may need help in the fields in the spring. We'll head back to Chicago next fall."

Stopping at business offices and talking on the street, he and Fred Sr. learned that a widowed farmer eight miles south of Le Mars had lost his helper, his son, who had married and moved away. The farmer needed help in the house and with the farm. He was looking for a partner.

"It won't hurt to talk to him," Fred Jr. told Miriam.

The trip to Chicago was abandoned. Fred Jr. went into the business of raising hogs on an Iowa farm, at $5 an acre as partner. Here in Le Mars was a railroad to Chicago. They could make money raising hogs, he was told. Chicago would wait.

The first year meant more hard work caring for hogs and bringing in the harvest. During the second year Fred and his father felt things would "look up." He was getting used to life in Iowa and he felt good about it.

Suddenly, hogs in the area began to sicken. An epidemic threatened. Special care was taken to keep hog pens clean and to nurse sick animals.

Fred Jr. was worried. Miriam saw the same signs she had seen in South Dakota.

"Miriam," Fred Jr. came into the house one day, sat down, his head bowed in his hands.

"Miriam, the hogs are dying. They're all dying. Of cholera."

Harold Wells was born in South Dakota

(Photo courtesy of Suzanne Wells Stocker)

The Leave Taking

Thought to be a congratulatory card following a birth announcement when Fred and Miriam's first child, Harold, was born in 1908

Chapter Seven

*Storytelling, oral history, is a part of continuity…
our universal hunger for stories cuts through… identities*

Studs Terkel

When the Wells family arrived in 1911, Le Mars boasted several banks and an Opera House, built in the late 1800s

(Photo courtesy of Plymouth County Historical Museum)

The Letter

Miriam stood at the sink washing breakfast dishes when her young son Harold came running into the house slamming the screen door behind him.

"Mama, Mama. The mailman just came. Here's a letter for you."

His mother wiped her hands on her big white apron and reached for the mail. The Sioux City Tribune and a letter. That was all.

The Tribune headlines grabbed her attention: "Sixteenth Amendment Authorizes Income Tax." Reading the first paragraph hurriedly, she rummaged through the silverware drawer for a kitchen knife, and prepared to slit open the letter. Probably from Jessica in Wellsburg. Wonder what's happening in Wellsburg.

Turning the letter over she noticed the postmark. The letter came from Chicago. Maybe sister Maggie had written.

Then she noticed the return address imprinted in heavy black ink…. It was from an attorney's office.

Why would I be getting a letter from an attorney? What's happened in Chicago? No more trouble I hope.

We don't want any more trouble. Maybe I should wait until Fred comes home. He can open it.

But Fred has been so worried. The hogs and the grocery bill… the $200 we owe. Fred is not used to this. What a mistake to leave Chicago. We always had money coming in.

He worked so hard to keep those hogs healthy. What will he say if there's bad news? We left Chicago not owing a soul. What can it be?

Miriam tapped the letter in the palm of her hand. She didn't want to open it.

Farming and raising hogs and corn is just not for Fred. He's trying to find something else…. Maybe work here at the dairy.

He worries about Roy too. Moving to Le Mars and this house. Fourth Avenue Southeast is a nice street. It's been a good change.

But Roy got so excited. He's only four. He ran from room to room to inspect our new place. So glad to get off the farm. Up and down stairs, up and down, so fast that he fell down the stairs. A hard fall… all the way down. Knocked him out. He just lay there like he was dead. I cried.

We thought for a while we had lost Roy. But he came to. And Doc said he would be all right. Roy is so full of ginger that it worries me.

Miriam tapped the letter against the table top and thought about these things. Should she open the letter? Or keep it until Fred came home? She walked toward a kitchen chair and sat down heavily.

No. She would not tell him about the letter. Not just

now. Not until later. Not until she had to.

Addressed to Mrs. Fred H. Wells, Jr. for delivery to Mrs. Miriam Ralston Wells, formerly of 253 W. Superior Street, Chicago the letter looked important. How had the post office found her? Why were there two names on the envelope?

Anxious now and impatient she tore at the flap and opened to a single thin sheet of stationery, addressing Mrs. Miriam Ralston Wells.

> *Dear Mrs. Wells:*
> *The Ralston Estate, that we have been named to handle, has been settled. Enclosed is your check for $900.*

Is this a joke, she thought. Who would do this? One of my brothers might. She felt the check, rubbed it between two fingers. Is it real? Will it cash? Or is this a trick?

She studied the check, hoping but afraid to hope. Written on heavy paper, it looked like a proper check.

I'll ask Fred.

"Fred, Fred... Where is Fred?"

The screen door slammed as Fred came in.

"Yes... yes.... Where are you Miriam?"

(Photos courtesy of Plymouth County Historical Museum)

Chapter Eight

How hard the battle goes, the day how long. Faint not, fight on! Tomorrow comes the song

Maltese D. Babcock

Fred Wells, Jr. in 1913. One school child remembered him delivering milk in a lined denim jacket and cap with Scotch ear flaps

(Photo courtesy of Wells Family)

The Beginning

"Never before have I ever had to borrow money… Miriam, I hope I never have to do it again."

Fred Jr., hand on the doorknob, hesitated. Opening the door he hitched up his overalls and moved outside. His step was slow, his broad shoulders sagged as he disappeared down the sidewalk.

"I'll be back… let you know what they say."

Will the bank loan us the money, Miriam wondered. Would they want collateral. The South Dakota land maybe. What other security could we give.

What will Fred do if the bank says no. The bank had been recommended. Downtown, the bank building looked safe and sturdy. Le Mars Savings Bank, that's where Fred was going. It's been there since the 1890s, Fred was told. This is 1913. That's twenty some years.

He needs around $250 to buy a milk wagon and get started. Go into the milk delivery business. He was in that business in Chicago. He knows it. He didn't know Iowa hogs.

Farmers in Plymouth County have a good many dairy cows. Fred can buy their milk, he figures, and deliver it in town to homes. And restaurants, maybe.

The $900 inheritance is almost gone to pay bills, Miriam thought. We need something to get us started… like a loan.

All morning Miriam went to the window, back and forth, watching for Fred. Harold was in school and would be coming home for lunch. Roy was playing quietly. She felt a little uneasy and wondered if she was in the family way.

Finally, Fred came in the kitchen door and called for Miriam "They're gonna let us know. Where are you? They're gonna let us know!"

He smiled, full of hope. Miriam's hope met his. His big, kind smile was one of the things that had attracted him to Miriam when she met him in Chicago. She smiled. The bank hadn't said no.

He was smiling when he told her later, "We're going to get that milk wagon and horse. When we buy that business from Bowers. The bank is going to let us have the money." Miriam's courage took a leap.

"Sears and Roebuck had those wagons for around $70 when we left Chicago. We made a good deal. I can go out and talk to farmers while we're waiting for the money. We're in business, Miriam!"

The large washroom in back of their home on Washington Street would make a good place for handling milk. Early in the morning, farmers could bring in milk… in big five gallon cans. The milk could go right into ice water.

Driving up and down each street in Le Mars in his delivery wagon, ringing doorbells, Fred could work up customers.

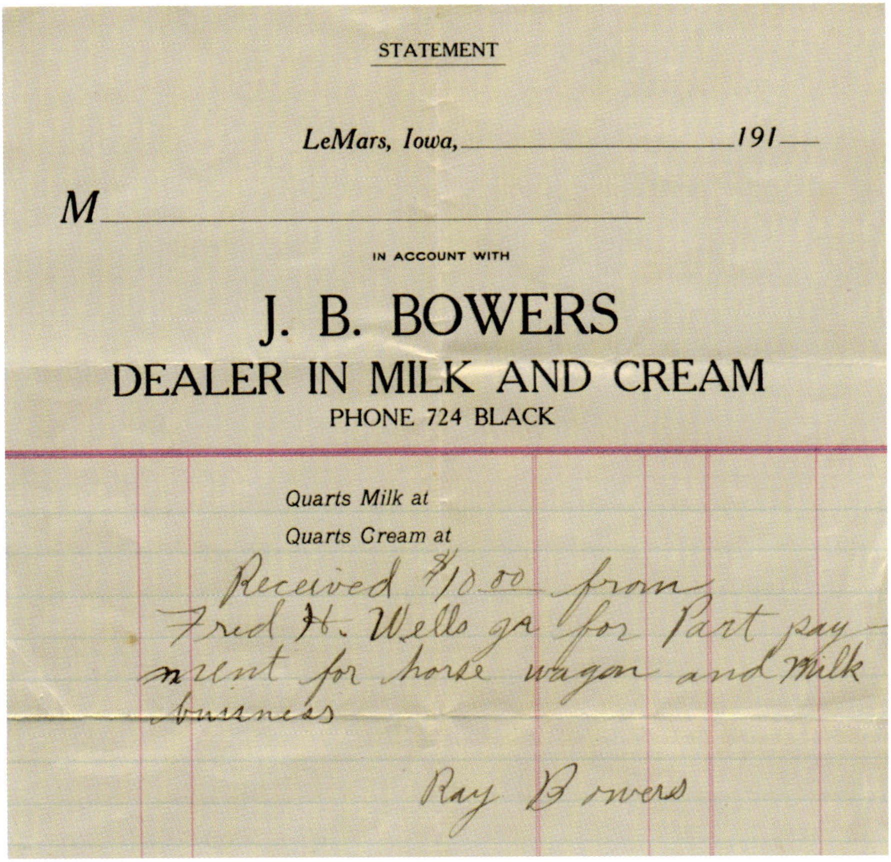

> **STATEMENT**
>
> LeMars, Iowa, _____ 191__
>
> M_____
>
> IN ACCOUNT WITH
>
> # J. B. BOWERS
> ## DEALER IN MILK AND CREAM
> PHONE 724 BLACK
>
> Quarts Milk at
> Quarts Cream at
>
> Received $10.00 from Fred H. Wells jr for Part payment for horse wagon and milk business
>
> Ray Bowers

In a short time Fred drove the streets each day with his delivery wagon bearing the Wells' Dairy name.

Housewives needing milk answered the door with a pan in their hands. Fred ladled milk into their waiting pans. Women liked having milk delivered to their doorstep.

Wells' Dairy milk business was building. Fred paid the rent, paid farmers for their milk, and the bank. He smiled every day of the seven long days that he worked each week.

Each evening Miriam greeted him at the door at day's end. She arose early to get breakfast for her family, get Harold

The Beginning

off to school, and help Fred with the milk.

And there were those trips up and down stairs, up and down to bring meals to Roy, urge her little boy to eat, bathe and change his bed clothes.

Roy had been ill. At night she got up each time she heard him murmur or cry out. He's so little to be so sick, Miriam thought as she studied his small figure in bed.

The doctor is coming today. I pray he can do something. Roy hasn't been downstairs for mealtime in quite a while.

When Fred came home, her news was not good.

"The doctor says that Roy has brain fever."

It would be a month before Roy was able to come downstairs to be with his family again.

Later, Fred Wells in a business suit. The delivery wagon has been painted a gleaming white and gold leaf lettering added to the Wells' logo

(Photo courtesy of Wells Family)

The Wells Spring

This contract made and entered into this 24th. day of October 1913 by and between Ray Bowers party of the first part and Fred Wells Jr party of the second part, both of Plymouth County, Iowa. Witnesseth.

That first party for and in consideration of $250.00 paid this day by second party hereby sells and transferes to said said second party the following personal property. One grey horse, one milk wagon, two barn cans, three 20 quart cans, sixty ½ pint jars, 60 pint jars and the good will of the milk business he has in city of LeMars, Iowa.

It is further agreed that Ray Bowers shall sell the milk from his cows, delivered on the place at which he now resides at the price of two cents a pound, said milk is to be furnished from not less than ten cows and not more than fifteen cows. Said milk to be cooled by first party, said second party to furnish the cooler for the milk.

Said milk that Ray Bowers is to furnish shall test not less than 3% and just as it comes from the cow, and he hereby agrees to furnish said milk until Mch. 1st. 1915.

Said second party agrees to pay for said milk every two weeks.

First party is to have the use of said horse and wagon until Nov. 1st. when said second party is to commence taking the milk.

First party agrees not to go into the milk delivery business in LeMars for a period of five years, providing said second party is operating a milk business in LeMars for said period of time.

...Ray Bowers......
...Fred H Wells Jr.........

The Beginning

OUR PASTEURIZATION IS YOUR PROTECTION

F. H. WELLS
PASTEURIZED
DAIRY PRODUCTS

RESIDENCE 115-2ND AVE. S. E.

_____ 192____

PHONE 711 BLACK

LeMars, Iowa

Chapter Nine

No mind is much employed upon the present; recollection and anticipation fill up almost all our moments

Samuel Johnson

Roy, baby Harry Lee "Mike" (front) and Harold (standing) have been combed and dressed in their finest for a 1915 picture

(Courtesy of Shirlee Wells)

Trip Back To Wellsburg

Two little boys in knee pants boarded the Northwestern and Dakota train at Brunsville. School had just dismissed for the Summer of 1917.

Two hearts pounded in anticipation of their trip. Harold clutched two tickets and handed them to the conductor. Roy inched close to his brother, standing as tall as he could to appear more grownup.

"Where you boys goin'?" The blue-coated conductor smiled, studying each one.

"To Philip." Harold was their spokesman.

"Where?"

"Philip… South Dakota."

"What for?"

"To see Grandma and Grandpa Wells… at Wellsburg."

"Why?"

"We're gonna help them this summer."

The all-day trip, exciting, nevertheless got a little tiresome by late afternoon. Miles and miles of prairie grass. All day long. Farm buildings once in a while. It all looked the same. A sod house sometimes. No trees. Some farm animals. Most smaller animals, when they heard the train whistling and grinding over the rails, ran for their lives to find cover. Wary whinnying horses shied, birds shrilled in anxiety and took to the air. Frightened chickens, the few they saw, squawked, hogs squealed. Prairie dogs disappeared down holes to their home.

Grandma and Grandpa Wells met the two at Philip's train station in a Model T Ford. The car, covered with dust kicked up on the long drive twenty-five miles from Philip, was black and shiny underneath. A spare tire nestled in a cup along the running board. Large, bold headlights gave the black radiator grill staring eyes. The windshield, sticking up above the steering wheel, made the whole car seem higher than a horse. It's rattle was a good noise after the train's thundering steam engine… a good noise and a change from the clip, clop of a horse and buggy.

Another long, jolting ride… then the Wells farmstead spread out before them, a bare-of-grass-farmyard, a barn, cattle shed, machine shed, corncrib, chicken coop.

The general store and post office in the over-large house had been shut down. The dance hall was in disuse. Homesteaders no longer came for Saturday night dances.

After the dust storms and dry years of 1911 most homesteaders "went broke." They picked up their meager belongings and left, looking for a better life elsewhere.

Fred Wells Sr. and Clara had gone to Iowa with Fred and Miriam but returned to Wellsburg two years later. Jessie, their oldest daughter, married to Philip Kenaley lived nearby. Ella Mae married Roy Hill. The two lived near Wellsburg.

Harry C., their oldest son, had gone to Doland in South Dakota, to homestead. The children were slowly going their own way. Ray and Howard to Chicago. Florence and Stella worked at farm homes in the Wellsburg area.

Nellie, the youngest, still at home, became manager of the 160-acre farm as Fred and Clara aged.

"I need a wire stretcher," Nellie told Roy after the boys had been there a few weeks. "Jesse and Philip have one. Harold is busy helping Grandpa. You can get it for me."

"Aunt Nellie saddled a horse," Roy says. "And I'm tying his strap to the fence at Jesse's when that horse bit me. In the chest!" Ruefully he notes the bite didn't hurt. Too much.

"That horse didn't like the noise of the wire stretcher across the saddle. When we got back at Grandpa's, I picked up the pulley and threw it off. The pulley and the chain fell on the ground."

The clanking chain spooked Roy's horse. "He ran for the barn. I looked at that barn door. It was low and I thought I better duck.

"I landed in the barn in the straw. My leg was scraped. And Grandma stood there bawling. They were all watching… scared.

"But that saved my life. When I ducked."

Threshing time in July also made the summer memorable. Roaring like a locomotive, a huge tractor and thresher pulled up into the Wells farmyard early one hot morning. Neighbors with teams and wagons fastened a long black belt into place between thresher and tractor. Dumping shocks of grain into the cavernous mouth of the thresher they watched as the separated grain poured into their wagons.

The last shock had been tossed into the machine when a rattlesnake coiling in the grain rattled a warning.

"I heard it." Roy says "I was close. Grandpa's dog stood right by my leg. That dog got it right in the hip.

"Grandma put a poultice on the dog. He was very sick for a few days. But he lived. That dog saved my life."

In late August before school started that fall, Fred Jr. and Miriam drove to Wellsburg from Le Mars, in their new Model T Ford. The Dairy was doing well enough for purchase of the Ford. In cash.

The Wells family had become one of hundreds to thousands of Americans saying goodbye to the stagecoach and horse and buggy when they purchased a "tin lizzy."

America was changing. If they had not seen a silent film, at least they had heard or read about the Keystone Kops, Charlie Chaplin and "The Great Train Robbery." Harold Lloyd, "Skinny the Rat," and Cecil B. DeMille were becoming household words.

"Mama," Roy told his mother just after Fred Jr. and Miriam arrived in Wellsburg to take them home, "Mama, I got awful sick… a terrible pain. But it went away."

Back in Le Mars, the pain returned. Roy was sick again.

"I couldn't stand a sheet touching me. I was sick five or six days. They called a chiropractor first. He didn't know what the trouble could be. Then they called Dr. Fettes."

"We'll operate. Today," Dr. Fettes said.

He found a ruptured appendix. Gangrene had set in. "I laid on my back for a month in the hospital. I wasn't supposed to move. Might break the incision. They didn't expect me to live. I fooled them."

The ride to Wellsburg and Philip had been the biggest train ride in Roy's young life, except for the time when he and Harold went to Chicago with Miriam to visit her sister, Maggie Zobel. That's when Miriam lost the diamond ring that Fred had given her.

In the train's washroom, washing her hands, she removed the ring. The ring fell into the sink and slipped down the drain.

It flushed down onto the tracks of the moving train below.

The ring was gone forever.

Her grief went on for days, Roy said. "What can I tell Fred?" she would ask.

"What will I say to Fred?"

Fred and Miriam Wells, Harold (left) *and Roy are in their best suits for this 1919 picture. Fred holds Harry Lee "Mike"*

(Photo courtesy of Suzanne Wells Stocker)

The Wellsburg farm was gradually improved until it looked like this before Fred Wells, Sr. and Clara left for Iowa a second time in 1923

(Photo courtesy of Doris Wells Zimmerman)

Roy was hospitalized here for a month as a child

(Photo courtesy of Plymouth County Historical Museum)

Trip Back to Wellsburg

Chapter Ten

When a person dies, he only appears to die. He is still very much alive in the past.... All moments, past, present and future, always have existed... always will exist

Kurt Vonnegut

Harold in 1932

(Photo courtesy of Suzanne Wells Stocker)

Harold, First-Born

Harold, on his way to classes, full of the pent-up energy of spring, kicked at a small winter-darkened piece of icy snow. He had talked with his father the night before about something that had been gnawing at him. He had made a decision.

Young maple trees along Central Avenue were budding, stirring within to throw their energy into full leaf. He was 17 and an eleventh grader at Le Mars High. On August 4 he would be 18. Spring sports were looming. His friends liked football and basketball. A good student, Harold wasn't interested in sports. Nor was he enthusiastic about studying.

Harold had decided to quit school. This would be his last day. Always cautious, the boy becoming man had spoken with Fred Jr., about going to work for him full time. He was eager to earn money and take the role of an adult.

The dairy business was growing. Delivering milk after school and on weekends… ever since he could lift a milk can he had helped… taught him something about the business. He liked it. And he knew there was "a lot of work to be done."

Studying the sidewalk, he lifted a stone with the toe of his shoe. Tomorrow, April 1, 1925 could not come soon enough. He stood thinking in front of the high school.

A good worker, he was careful about the way he treated Wells' Dairy customers. He had shown interest and an aptitude. His father was a patient teacher.

The oldest of Fred Jr.'s four sons, responsible, Harold had emerged from the boy who held their train tickets when Roy and he visited Wellsburg and their grandparents in 1917. Feeling his way slowly, Harold followed a respectful distance now to his father's self-assured, confident lead.

Known by his siblings as the anxious "worrier," he was not a risk-taker. In his new role at the dairy, he nevertheless moved from making deliveries of ice cream and milk to Le Mars homes and businesses to becoming the first ice cream salesman on routes outside Le Mars, and manager of the delivery system.

Harold was the "pessimist," however, while Roy was the "optimist." Roy could say with enthusiasm:

"Boy, business is so good, we're making tremendous ice cream."

On the other hand, Harold, counting the cost, replied, "Yes, but equipment is wearing out awful fast. We're using up the ice cream freezers."

During the 1930s, Roy tells, "Harold was sure we were going broke." Times were tough.

"We'd better find a way to pay the bank," Harold told his father.

Fred had a ready answer. "It's already paid off."

Was it an early experience as a child that thrust Harold into the mold of one trying to keep out of harm's way?

Roy, the keeper of memories, recounts the story he often heard his father tell. When Roy and Harold were still kids on the Le Mars farm, the two boys were playing in the hay barn.

"The block and tackle to raise the hay was hanging down," he says. The boys swung on it, riding with it up and down, watching it spring out of their control.

"Then we left, leaving the block and tackle hanging down. Too low.

"Dad's partner had just bought a brand new buggy." He came riding into the hay barn in his new horse and buggy.

That block and tackle "tore the buggy top up. He was mad."

An angry farmer, distressed over a new, destroyed buggy top must have imprinted deeply on the heart of a child, and the child's sense of responsibility and fear over loss.

Whatever the reason, Harold became increasingly prudent as he grew older.

His brothers speak of Harold as a "wonderful man" and a hard worker, thoughtful and competent.

In 1933 Harold married Bernadette "Betty" Koppes a young woman from a "very good" family in Le Mars.

Beset with health problems, the young family became another source of concern for Harold. The older son, Freddy, died young.

The younger son, Ronald, joined the U.S. Marines and suffered service-related difficulties that derived from the 1960s.

He lives in Des Moines, after a career at Firestone Tire and Rubber Company. Suzanne, Madonna and Rita, Harold's daughters, live scattered across country.

As manager of the delivery system, Harold bought new dairy trucks as needed, and finally as the business continued to grow, maintained 90 trucks used as delivery vehicles in four states.

It was Harold who designed the body and refrigeration unit for the dairy's first truck that displaced the old ice and salt trucks.

It was Harold who kept a watchful eye on the plant during weekends.

"We didn't have automatic controls on the tanks at that time," Fay, the fourth son tells. "Ice cream mix was held in the tanks over the weekend. We'd go in and open the ammonia valves… cool the mix and shut the valves.

"Harold was at the plant, shutting the valves one Sunday morning. He was found collapsed on the floor in one of the rooms." He never regained consciousness.

He was 63. He had had a massive stroke.

When his funeral cortege moved slowly along Central Avenue the once-young maple trees arched over his procession, clasping hands that spring day in March, 1972.

Harold's grandchildren are: Donald James Stocker, Jr. and Ann Stocker Vatch, children of Suzanne; Tiffany Searing, daughter of Madonna; Eric J. Tadlock and Angela Lynn Tadlock, children of Rita.

A growing family Harold, Mike and Roy, (standing) Fred Jr., Fay and Miriam

(Photo courtesy of Roy Wells)

Harold, First-Born

Harold in later years is shown (right) with his family: his wife, Betty holding youngest daughter Rita; son, Fred with his sister Madonna and (standing) Ronald and Suzanne

(Photo courtesy of Suzanne Wells Stocker)

Harold, Fred Jr., and Roy are photoed at Sioux City's Martin Hotel during a meeting

(Photo courtesy of Wells' Dairy)

Harold, First-Born

Harold Wells, 64, dies Sunday; partner in Wells Blue Bunny Dairy

Harold R. Wells, 64, a partner in Wells dairy here, died unexpectedly Sunday night at Floyd Valley hospital.

Funeral services will be Wednesday at 10:30 a.m. at St. James Catholic church with Rev. John Turza officiating.

Burial will be in Calvary cemetery under direction of the Luedtke funeral home.

Friends may call at the funeral home after 2 p.m. Tuesday. Prayer services will be at 8 p.m. Tuesday at Luedtke funeral home.

Mr. Wells was found in the boiler room at the Wells plant Sunday about 6:30 a.m. where he apparently had been stricken with a stroke. He was taken to Floyd Valley hospital by Luken-Johnson ambulance.

At the time of his death, Mr. Wells was the purchaser of all trucks and cars and was in charge of maintenance of all the fleet for Wells Dairy.

Mr. Wells formed a partnership in the dairy with his father, Fred H. and brother Roy in 1935. Previous to that his father Fred H. Wells Jr. owned the plant and Roy

Harold R. Wells

and Harold were employed there. The senior Wells started the dairy in 1913 when he purchased a Le Mars milk route from Ray Bowers.

Harold is the oldest of four brothers who are all in partnership in the dairy. They are Roy, Harry and Fay. The fifth partner is a cousin Fred D. Wells of Sioux City.

Harold Wells was born Aug. 2, 1907, at Wellsburg, S.D., a son of the late Fred H. and Mary Ralston Wells Jr. Wellsburg was founded by Mr. Wells' grandfather, the late Fred H. Wells Sr.

Harold Wells and Bernadette Kappes were married June 9, 1932, at Ashton. Since their marriage the couple has resided in Le Mars.

He was a member of the fraternal orders of Elks and Eagles.

Mr. Wells was preceded in death by his parents.

Survivors are his wife, Bernadette of Le Mars; two sons, Fred of Le Mars and Ronald of Des Moines; three daughters, Mrs. Robert (Madonna) Serring of Salt Lake City, Utah, Mrs. Donald (Susan) Stocker of Rockford, Ill., and Mrs. Dan (Rita) Tadlock of Sioux City; three brothers, Roy, Harry and Fay, all of Le Mars and four grandchildren.

Chapter Eleven

If the very old will remember, the very young will listen

Chief Dan George

Roy in 1929

(Photo courtesy of Shirlee Wells)

Roy, Second Son

"Why did you hang up, Dad?" Roy noticed that his father seemed unusually agitated.

Fred Jr. sat down at the dinner table and shook his head in disgust. "Some crazy fella," he said, "trying to pull a trick." With a large spoon he scooped from a steaming hot bowl of mashed potatoes to his plate.

"Who was it?" Miriam turned to her husband. Busy at the stove she had not heard his telephone conversation. Since she most often answered for the dairy when the phone rang she was more than curious.

"Some guy who said he was from Kansas City," Fred told her. "Making promises. Money. Crazy."

The phone rang again, Miriam pulled a pot off the stove and set it aside. Reaching for the phone she asked, "Who?" Then she listened intently.

Before giving the caller information that made Fred Jr. scratch his head, Miriam asked more questions. Grabbing a pencil she made little notes on a phone pad. Then she hung up.

"They want the names and addresses of all your sisters and brothers. Birth dates and so on. Someone on the Wells side died in Chicago. You're going to inherit some money."

The phone call that Fred Jr. found less than believable netted the family $450. Who died, where or when has been lost to history.

"We weren't hard up then," Roy remembers. The dairy had been prospering in the 1920s. "Both parents' health was good. Dad loved to travel. He began to leave things for us boys, Harold and me, to run."

Roy's experience with the dairy business had been extensive. He helped from the time he was in second grade. "Mother bottled the milk. She had a two-gallon can with a spout... five or six inches long. She'd pour the milk into quart bottles. I put caps on the bottles... when I was seven or eight...."

Slowly the Wells family fortune improved. The dairy was expanding. "Before the 1920s it was difficult... the worst times.

"A quonset hut built in back of the house at 115 Second Avenue became a new dairy building. The house was moved across the street," Fred Jr. then built a new home for the family.

Franklin School, near the new house, Roy said, "had a great hill for sleds and toboggans. Drew big crowds. They'd carry buckets of water there to pour down and get solid ice on that hill. The street was blocked off. They'd sled clear to the railroad tracks.

"Maurice Hoorneman lived nearby, across from Plymouth County Jail. Some fellow driving through the blockade... killed Mr. Winters, Maurice Hoorneman's father-in-law. The families were close…played bridge together."

For summer recreation "we walked out to Grimes Sandpit… swam to the raft and across the sandpit. We liked to go to shows. Shows at the Royal Theatre and The Pix cost ten cents."

Times change. The old jail and Plymouth County Courthouse are still there, both listed on the National Register of Historic Places. Franklin School has been displaced with a new vocational school.

"But I didn't like school," Roy said. "I had been delivering milk before and after class. In the spring of ninth grade I quit. Dad had added ice cream to the dairy business. I was needed. That was my first full-time job."

The work ethic was instilled early in young Roy.

"I started freezing the ice cream with a small machine and dasher. Ten gallons at a time. For three years I did it. Biggest trouble, we needed two or three machines to keep up.

"There were two steps… put it together, pasteurize it. And it went through a freezer, pumped through a pipe at about twenty degrees."

On a Sunday, November 14, 1931 Roy married Eunice "Kitty" Lundgren at a ceremony "in a brand new house we rented. My folks and Fay and her folks, Knute and Alta Lundgren, were there." Roy was 22.

Before leaving on their honeymoon the newlyweds stopped to see his grandparents, Fred Sr. and Clara on Plymouth and Second Street SW. The elder Wells lived there with Roy's Aunt Nellie.

"Grandma was dying of cancer." Clara Wells died two months later, January 16, 1932.

"In 1936 we installed a continuous ice cream freezer. It turned out 150 gallons an hour and made a smoother, better product. Two more employees were added." Business grew slowly.

By this time Roy had become general manager and manager of production, without the title… buying all supplies and ordering, among other duties.

"When war broke out in 1941 we lost most of our production people." Roy was deferred since he was in a food industry.

"But," he says, "I was ready to go."

During this time, two daughters, Darleen and Shirlee were born, and a son, Steven (who now heads a Wells regional office in Florida).

Roy had worked at the Sioux City plant for several years. In 1954 Fred Jr. died, leaving full responsibility to the sons.

By the 1960s and 70s Roy had moved from general manager and manager of production without the title, then to vice president and treasurer in 1977 upon incorporation of Wells' Dairy.

In the 1980s he retired to spend time in Florida with his growing family and his Chris Craft at Pompano Beach.

Earlier he boated at Lake Okoboji in Northwest Iowa.

Now winters were spent in Florida enjoying his boating

hobby and cruising in his small luxury yacht "all the canals on Pompano Beach." During his boating years he could be found behind the wheel of three different boats, one with "two huge diesels."

Each March he returned to Iowa "to do income tax." He and Kitty spent summers in Iowa.

Kitty came up from Florida to be with Roy in Le Mars in June, 1996. In August she became ill. Following a difficult time with pneumonia, she died. Thus ended a marriage of nearly 65 years. Her death shattered Roy's plans for more winters with Kitty in Florida.

Their daughters closed the Florida home. Roy grieved with them in Iowa during a sad year.

The family Chris Craft explores Florida waters now with Steven Wells at the helm. He lives at Pompano Beach.

Steven Roy Wells, at 14, with a permit, began work at Wells' Dairy part-time after school and weekends. Finishing his education in Health and Business, he worked for Wells' until 1978. Then as Regional Director for Fort Lauderdale Hospital, he was overseer of six East Coast hospitals. "Tired of travel and missing the dairy," he went back to Wells' in 1992. Steven is Regional Sales Manager responsible for Wells' sales representatives and retail ice cream sales in Florida. He is a member of Florida Dairy Association. His hobby is boating Florida water in the family yacht.

Roy's grandchildren include Daniel Renken, Julie Renken Jacob, David Renken, Tom Renken and Joni Renken Clemens, children of Darleen; Dr. John Clemens and Cheri Clemens Rieken, children of Shirlee. Roy has fifteen great-grandchildren.

Roy, Second Son

At Long Beach, California in 1939, Roy caught a bundle of fish. Kitty's family lived on the West Coast. Visits to California were frequent

Steve Wells (below) boats the Florida waters in the family yacht

(Photo courtesy of Steve Wells)

Kitty and Roy Wells pose in 1950 for a picture taken in their home. Darleen (left) and Shirlee now have a small brother, Steve

(Photos courtesy of Shirlee Wells)

Wells' employees celebrate a festive Christmas party at Le Mars Country Club. Mike and Frances Wells (left) sit across the table from Mr. and Mrs. Keith Wilcoxen. Roy and Kitty are at right

(Photo courtesy of Wells' Dairy)

Roy Wells was elected president of Association of Ice Cream Manufacturers of Iowa in January, 1944. This picture was taken at Hotel Fort Des Moines and appeared in the Des Moines Register January 20, 1944, with a two-column story about the important meeting

(Photo courtesy of Shirlee Wells)

CELEBRATE GOLDEN WEDDING TUESDAY

Mr. and Mrs. F. H. Wells sr. Observe Fifty Years of Wedded Life

Mr. and Mrs. Fred H. Wells sr., 122 Plymouth street, will celebrate their golden wedding anniversay tomorrow at the Odd Fellows hall.

Their entire family consisting of five daughters and five sons are most happy to be with them. The children are:

Mr. and Mrs. Harry C. Wells and family of Sioux City; Mr. and Mrs. F. H. Wells, jr., and family of LeMars; Mrs. Philip Kensley of Seattle, Wash.; Mr. and Mrs. Ray S. Wells and family of Spencer; Mr. and Mrs. Howard I. Wells and family of Chicago; Mr. and Mrs. Roy Hil land family of Sioux City; Mr. and Mrs. Charles Schmidt and family of Wenatchee, Wash.; Mr. and Mrs. William Hampton and family of Martin, S. D.; Miss Nellie Wells of LeMars.

The guests will include: Mrs. B. Schmidt, LeMars; Ms. M. Alfred, LeMars; Mr. Zobel, Chicago; Mr. and Mrs. Freiburg and son of Havelock, Ia.; Mr. and Mrs. H. Shelzien and family of Gilmore City, Ia.

The table and hall decorations will be of gold and white. Yellow and white carnations will decorate the tabe.

Al Spaan of Ames, who lived in LeMars about two years ago, is visiting friends and relatives in LeMars today.

DEATH CLAIMS OLD RESIDENT

Mrs. F. H. Wells, of This City, Dies After Lingering Illness

LEAVES LARGE FAMILY

Celebrated Golden Wedding a Few Months Ago

Death claimed Mrs. Fred H. Wells, residing at 122 Plymouth Street SW., Saturday morning, January 16, following a lingering illness caused by an internal malady.

Mrs. Wells was born May 25, 1858, at Bridgeport, Connecticut. She lived there during her childhood and young womanhood. At the age of eighteen she moved with her parents to Chicago and was united in marriage with Fred H. Wells July 3, 1881, at Chicago. To this union nine children were born: Harry C. Wells, Sioux City; Fred H. Wells, Jr., Le Mars; Mrs. Jessie C. Kensley, Seattle, Wash.; Ray S. Wells, Spencer, Iowa; Howard I. Wells, Chicago, Illinois; Mrs. Roy R. Hill, Sioux City; Mrs. Chas. Schmidt, Seattle, Wash.; Mrs. Will Hampton, Martin, S. D.; Miss Nellie Wells, Le Mars.

Move to Western Dakota

After living in Chicago for thirty years the Wells family moved to South Dakota. For the past twelve years they have lived in Le Mars.

Mr. and Mrs. F. H. Wells celebrated their fiftieth wedding anniversary in this city with all their children and their families present on the occasion. Shortly after the celebration of their golden wedding Mr. and Mrs. Wells left on a trip in the East to visit their old home in Connecticut and in Pennsylvania. Since their return from their trip Mrs. Wells had been in poor health and failed continually until the time of her death Saturday morning at 10 o'clock. She was 73 years, 7 months and 22 days old.

Mrs. Wells when a girl became a member of the Episcopal church. She was esteemed by all those who enjoyed her acquaintance and made many friends in Le Mars since coming here to reside.

Funeral Held Monday

The funeral was held Monday afternoon from the home of her son, F. H. Wells, 115 Second Avenue SE., at 2 o'clock and services held in the First Methodist church with Rev. F. Earl Burgess conducting the last rites.

Death Claims Fred H. Wells

Resident in Le Mars Twenty-four Years

Fred H. Wells, 77, a resident of Le Mars, for twenty-four years, died Sunday afternoon at his residence, 122 Plymouth street SE. Mr. Wells had been in failing in health for a long time. (April, 1939)

Mr. Wells was born November 22, 1861, in Erie county, Pennsylvania. At the age of four years he was left an orphan and made his home with an uncle, John Cole, until he was sixteen years old.

When a young man he lived in Chicago and was married there in 1881 to Clara Marvin. They were the parents of nine children.

He is survived by H.C. Wells, of Sioux City, F.H. Wells of Le Mars, Mrs. Philip Kensley, Mrs. Roy Hill, Mrs. Charles Schmidt, Mrs. Will Hampton of Seattle, Wash., Howard Wells of Chicago, and Mrs. Roy Chenhall of Le Mars. His wife and one son, Ray Wells, proceeded him in death.

Mr. Wells, who was engaged in business in Chicago went in 1907 to western South Dakota and came to Le Mars in 1915. He was engaged in real estate and insurance business among his other business enterprises.

The funeral will be held this Tuesday afternoon at 2:30 p.m. at the first Methodist church, Rev. W. M. Hubbard officiating, and interment made in the city cemetery.

Luken's funeral home in charge of arrangements.

Chapter Twelve

Stories last longer than the people who live them

Wallace Stegner

Harry Lee "Mike" Wells, 1939

(Photo courtesy of Shirlee Wells)

Harry Lee aka 'Mike'

Mike, christened Harry Lee Wells, moved into Fred Jr. and Miriam's life on a cold day, December 15, 1913, at Le Mars Hospital.

Miriam had given birth to Harold and Roy at home in Wellsburg, twenty-five miles or more from a town, and "eighty-five miles from a railroad" as Aunt Nellie Wells wrote years later.

Harry Lee, always known by his nickname, "Mike," made little impact at birth upon his two older brothers.

Until Mike was "seven, maybe eight or nine," when it happened. Roy remembers it well.

"Have you seen Mike?" Miriam asked his brothers.

Mike was missing. Last seen with his friend, Bob Duster, a neighbor boy, he hadn't checked in with his mother for several hours.

"Where is he?" Fred kept asking. "Where is he?"

Miriam phoned neighbors and friends. "Have you seen Mike or Bob?" Bob's parents hadn't seen the two boys, didn't know where they were.

No one had seen them.

Mike didn't like school. Perhaps he had run away so he could avoid school. His father had engaged a tutor to help Mike with hated schoolwork. Still Mike resisted school.

"Maybe the boys have gone to Remsen," Bob's parents suggested, "to see Bob's relatives."

Miriam and Fred drove to Remsen.

No one had seen two boys in little hurry to get anywhere, much less home.

Roy and Harold were sent out to search. Le Mars police were alerted. Everyone up and down the street was queried again.

Finally two small boys came trudging home.

"Where have you been? Where?"

"To Remsen."

"We went there. Couldn't find you. We were worried!"

"We walked… the tracks."

"Oh! You might have been killed. What if a train had come along? And you were on a trestle?"

Their parents were happy to have the boys back safely. They didn't scold. "Did you meet any tramps?" friends asked.

Mike spent little time with Harold and Roy. His interests did not parallel theirs. Since the dairy was doing well Fred did not ask Mike to quit school and work in the business as his older brothers had.

In high school, Mike loved football, was captain of the team.

"He was a great football player," Roy says. "I can still see him… a good runner… good at getting the guy with the ball."

Mike was graduated from Le Mars High School with honors, class of 1934, was class president, and a member of National Honor Society. During their high school years he had been dating Frances Marx, a Le Mars girl. Frances went to Sioux City to work after graduation.

Mike followed and got a job with Roberts Dairy. The two were married in Sioux City in a quiet ceremony June 22, 1935.

"By themselves. Mike was the quiet one. He told no one anything," Roy says.

Their marriage in the middle of the Great Depression meant a small wedding.

Life continued to be crowded with work hours. "He was always running errands (for Wells' Dairy) while he was in high school," Frances Marx Wells says. "Taking the truck to deliver milk.

"Later he hauled milk to Sioux City. He had developed a milk route in the residential district… and to stores. And he went around collecting (for the dairy) too. He worked long, long days. He worked really hard."

Frances was employed at Home Owners Loan and for Swift and Company. Their first child, Robert James, was born in 1943 in Sioux City.

Before David Michael's birth in 1947, they were back in Le Mars where Mike, who had shown an unusual capability for anything mechanical, had risen in engineering jobs at the dairy and later became head plant engineer.

He not only handled maintenance problems for a growing company, but also maintenance personnel and took charge of

Harry Lee aka 'Mike'

buying maintenance equipment. He was a refrigeration expert.

"He was very good with mechanical problems," Roy tells. Responsibilities in production were shared with Roy.

Two of Mike's good friends, Francis Nemmers, a plumber and Bob Wright, an electrician, formed a long-time bond, bound together by similar interests.

Mike took early retirement from maintenance and production in 1975. He and Frances spent time at their home on Lake of the Woods.

Their son David became manager of the South and North ice cream plants. Robert lives in Reno, Nevada and has been "involved in things mechanical," in his own metal recycling and fix-it business.

After suffering a broken hip, Mike at 82 was moved in 1996 to a new area for special cases at Happy Siesta Rest Home in Remsen, the Remsen he had liked as a child.

Mike died in Remsen, Thursday, February 19, 1998. He was 84.

The grandchildren of Harry Lee "Mike" and Frances Wells are Michael T., Brian D. and Neal J., the children of David Wells. Bradey Michael Wells and Tyler David Wells are great grandsons.

It was a proud day in 1933 when Mike posed with his first automobile following graduation from Le Mars High School

And a proud day when Mike and Frances Marx were married in Sioux City June 22, 1935

(Photos courtesy of Frances Marx Wells)

Harry Lee aka 'Mike'

Mike was in the U.S. Navy, Seabees Division, when he came home in 1944 for a visit and photograph with Robert, his first child

(Photos courtesy of Frances Marx Wells)

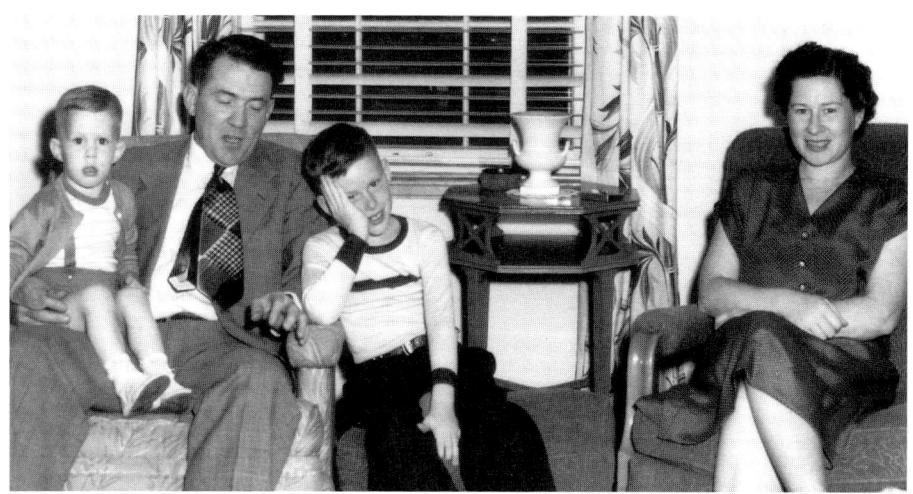

Mike holds son David while Bob and Frances, his wife, sit for a 1950 family portrait in the home of Roy and Kitty Wells

Harry "Mike" Wells (left), Fred Wells, Jr., Fay Wells and Roy Wells, early 1940s

(Photo courtesy Wells Family)

Harry Lee aka 'Mike'

Former dairy executive dies

Harry Lee "Mike" Wells, a former Wells' Dairy manager and part-owner, died Thursday at the Happy Siesta Health Care Center in Remsen at the age of 84.

Wells' career in the dairy industry began when he worked with his three brothers, Roy, Fay and Harold, in the family dairy while he was still attending Le Mars Central School.

His father, Fred H. Wells, founded Wells' Dairy with a milk distribution route in Le Mars in 1913, according to a Wells' Dairy corporation history. Today, Wells' Dairy is the largest family-owned dairy in the nation.

After graduation from Le Mars Central High School in 1934, he moved to Sioux City and developed a milk route, delivering milk door to door. Wells returned to Le Mars in 1940 to work at the dairy.

HARRY 'MIKE' WELLS

In 1943, he served in the Seabee's Division of the U.S. Navy during World War II and was honorably discharged in 1945, when he returned to Le Mars to continue working in the family business.

He served as head plant engineer and production manager responsible for the selection, installation and maintenance of all machinery and equipment. He was also responsible for the personnel of the engineering and production departments. He retired in 1976.

Chapter Thirteen

Compensation for loss of childhood and youth is the precious memories that remain

Anonymous

First Wells' Dairy building at 115 2nd Avenue SE housed an office and dairy behind the Wells' home. Taken about 1928 Harold stands beside his mother, Miriam, to his left, and young Fay in front of Miriam. (Others are unidentified)

(Photo courtesy of Wells' Dairy)

Fun Times

Three boys becoming men and a boy-child were sitting on their front porch at Second Avenue SE after a day of hard work. In this summer of 1928, early evening, they were looking for fun.

Fay, the youngest, about five, sat listening to the older boys, Harold, 19, Roy, 18 and Mike, 14.

"I looked up to them as adults," Fay said later, "not as playmates."

But the young men were in a playful mood. One pointed to a broom standing in the corner nearby. Miriam, their mother, like her neighbors swept daily.

"You know, Fay, you could straddle that broom," his brother said, "and fly." He sounded convincing.

"Yeah," said another, "just run and jump off the porch."

"Be sure to keep running," said the third.

A kid is always eager to try something new. This sounded like a big adventure.

"I got on the broom," Fay tells, "and I kept on running."

Rising up off the dirt and grass, Fay came back up the porch steps, dragging a broom and rubbing his face.

He faced his teasers, downcast. "Well," he said, "that didn't work."

"They were laughing their heads off," Fay recalls years later.

On May 20, 1923 Fay was born in the house at 115 Second Avenue SE. The location is now known as the North Ice Cream Plant.

"That old house was moved away two blocks east and sold. I remember Dad taking me through a new house they were building. The dairy was in back of that house.

"It was a few steps to the dairy office. We all had supper together and dinner at noon. If a customer came (while we were eating) I'd jump up and trot over seventy five feet and wait on them, ring it up on the old cash register and finish dinner. I was eight or nine then."

When Fay was a baby his grandparents, Fred Sr. and Clara Wells moved to Le Mars in retirement from Wellsburg, South Dakota.

"I remember Grandma cooking on a wood and cob stove in a house where they lived on Plymouth Street (west) close to downtown. They rented and lived in the back.

"Aunt Nellie Wells had a beauty parlor with big windows in front. It was fun to play in the beauty parlor, sit on the chairs and under the hair dryer. There were two big advertising signs in front to hide the yard from the street.

"They had holiday dinners there, parties in the back yard. A wedding once, with Chinese lanterns strung across the yard."

Grandparents Fred Sr. and Clara Wells

In 1931 Fred Sr. and Clara Wells celebrated their fiftieth wedding anniversary with all their children (left, front) **Stella, Fred Jr., Clara** *(wearing a corsage),* **Fred Sr. and Nellie;** *(back row left)* **Jesse, Howard, Ella Mae, Harry C., Florence and Ray**

(Photos courtesy Wells Family)

Fun Times

Survival was the key word during those times. Little interfered with work.

"Dad worked hard, seven days a week. He never got in the habit of being a consistent church goer.

"But Mother wanted to go to church. She started in Chicago when they were young. She and I would go together to the Presbyterian Church. We were always Presbyterian."

Despite the stringent work ethic that Fred Jr. clung to all his life, he nevertheless gave his family good memories of fun times together.

"Later, when the business began to improve," Fay tells, "while I was still a kid, Dad and Mother traveled a lot," even though Miriam did not enjoy travel as much as Fred.

"When they traveled, I'd stay with Roy and Kitty. But one time I went to Canada with them to fish for lake trout. Dad let me drive and sit in the front seat… there were long stretches of road.

"We portaged our supplies from one lake to another. One lake was 80 feet deep or more. We used long lines. When you'd get a bite on all that line… it was an experience!"

Fay easily grew into the work habits of his father.

"The summer I was 12," he tells, "I said I'd like to work in the dairy, and make some money. Roy put me to work on the mix maker. We made 200-gallon batches of ice cream mix. Dad paid me $3 cash each week for six days of work."

Although hard workers, family members always seemed to know how to have fun.

"I got to know the guys. It was a lot of fun. Right up

through high school I was the assistant mix-maker each summer.

"Roy would order rail car loads of 100-pound bags of sugar.

"Two of us dumped 10-gallon cans of milk and cream into the mix tank. (Now we pump it through stainless steel pipelines.)

"We'd carry all the dry ingredients from the warehouse to make the mix. We didn't have carts. We enjoyed that… carrying it by hand.

"I was in high school then. A lot of my friends were farm boys who worked in summer. Farm boys were hard workers. We didn't go to summer camp. We worked. That was our fun."

By the time Fay moved out of young adulthood the 1940s approached. A developing war over-shadowed the needs of a growing dairy business. The year Fay turned 18 the U.S. declared war on Japan and the Axis.

Wells men always have fun along with their work. At a 1962 convention, Fay, left, and Roy were pictured

(Photo courtesy of Wells' Dairy)

Fun Times

Fay Wells (left to right), *Harold Wells, Harry "Mike" Wells,* (center),
*Roy and Fred Dale Wells were snapped in front of the Le Mars Milk Plant.
Note milk silos in the background*

(Photo courtesy of Wells' Dairy)

Chapter Fourteen

We live historically and don't perceive it

Richard Eder in Los Angeles Times

Harry C. Wells posed for a formal portrait

(Photo courtesy of Fred Dale Wells)

Sioux City And Harry C.

Harry Cole Wells became discontented with life at Doland, South Dakota. Ambitious, he had tried farming, the hardware business in Doland and then real estate. He became a brakeman for Chicago Northwestern Railroad. Now he looked for something more profitable.

Brother of Fred Wells, Jr., Harry C. left Wellsburg, South Dakota, for Doland during the dry years that forced Fred Jr. to flee with his family in 1911.

Harry married Freda Levesen, daughter of a successful Doland farmer. They had two small daughters. Another child was expected.

Driven by a need to improve life for himself and his growing family, he contacted his brother Fred in Le Mars.

The brothers sat down together and came up with a plan. They would open an operation in Sioux City. Harry C. would be in charge.

Wells' Dairy in Le Mars had begun to make ice cream a

few years before after starting the business in 1913. Thirteen years of intense struggle were beginning to ease. By 1926 ice cream sales had helped boost the company. Sioux City, the two reasoned, should be a good new market for ice cream sales.

Since Wells' Dairy put its profit into the Le Mars operation, two partners were taken in for the Sioux City venture. Robert Harris, an ice cream maker at the Le Mars plant, and A. Paige, who worked there in production, each put $5,000 into the Sioux City business.

"And Harry did a lot of work," says Roy.

"This is good ice cream," Sioux Cityians began to report. Word got around. The product eclipsed that of other competitors.

"Ours, made of sweet cream, good quality, outsold the competition," Roy tells. "It went over big.

"They (competition) were in the butter business. They put their best cream into butter. They got a good price for it. The poor cream went into ice cream." Some said they used sour cream in making their product.

"They couldn't take it … (competition with Wells.)"

Two years later, Fairmont Creamery in Sioux City, Wells' chief competitor, said "We'll buy you out."

In 1929 Fairmont Creamery purchased the Sioux City plant.

"Harris and Paige made some money," Roy tells with a twinkle in his eye.

Wells' Dairy agreed it would not sell in Sioux City territory for five years.

"That sale took us right through the Depression," Roy says. "We didn't owe any money."

The Sioux City plant sold, Harry Cole began work at Fairmont's Sioux City plant adding to his experience.

When the five years of Fairmont's contract expired, "Harry was back on the street in Sioux City. We expanded in three directions," north to Craig, Brunsville, Maurice, Rock Valley, east to Remsen and Cherokee, and west to Akron. Wells' Dairy continued to produce a superior product.

The Fairmont people were "not happy" about the return of their competition.

"Fairmont threatened to sue, but they had no grounds for a suit," Fred Jr. was told.

And Wells' kept growing. "We plugged along," Roy recalls. "When World War II broke out, we were short-handed." Roy was deferred for six-month periods. "Milk was needed during the war by U.S. troops."

His brothers, Fay and "Mike" (Harry Lee), were drafted. Mike went into the Navy, serving in the South Pacific. Fay, in the Army Air Corps, was sent to the European theatre.

During the war, on June 22, 1943, Harry Cole Wells died after a long struggle with cancer... leaving a void in the Sioux City operation.

He also left a grieving widow, Frieda, two young daughters, Wilma and Marvel Marie, and sixteen year-old Fred Dale Wells.

Sioux City and Harry C.

Harry Cole Wells looked handsome in his uniform as a brakeman on the Chicago Northwestern Railroad during the early 1900s

(Photo courtesy of Fred Dale Wells)

Ray Kirby worked under Harry C. Wells during the Sioux City days

(Photo courtesy of Wells' Dairy)

Sioux City and Le Mars employees gather in 1948 for a company portrait. Front row, *(left to right)*, are Mike, Fay, Miriam and Fred Jr., Roy, Harold and Keith Wilcoxson. Back row second from left are Bill and Lloyd Grossenheider. Leon "Ole" Van Goor is under the Quality Check banner and the check mark. At extreme right, *(top)*, is Rich Koerner. (Others unidentified)

(Photo from Wells' Archives)

Sioux City and Harry C.

Chapter Fifteen

To tell and to read them (family stories) is to celebrate life

Anonymous

Typically farmers placed filled milk cans out at the road with other farm produce for pickup

(Photo by Robert Skidmore)

The Midnight Ride

A dozen five-gallon milk cans rattled in the panel bed as the Ford's tires grabbed the gravel in haste. Near midnight on County Road 60, it was dark.

At the wheel, Roy peered ahead and stepped on the gas. He had just spotted flashing lanterns ahead at the bridge. Roy knew what might be advancing toward him. Fred Jr. had warned there might be trouble that night.

Roy, Mike and Ole Van Goor finished their last milk pickup at a farm down the road. They were headed back to town.

Farm Holiday Association members were known to have plans to stop Wells' Dairy from buying milk from farmers. They had demanded that farmers dump their milk rather than accept plummeting prices lower than 25 cents a gallon. It was September and 1932 had been a tempestuous time. The Great Depression was churning through the Plains States.

Lanterns at the bridge waved wildly. The panel truck was commanded to stop.

Not to be intimidated, Roy was bent on carrying out his

responsibility. At age 23 he relished the challenge.

The lanterns kept flashing. The truck charged on.

"It's an ambush, Mike." Roy shouted above the Ford's engine and the gravel's crunch. "Get back to the cans." The truck veered a little left on the slippery gravel. "Don't let those cans fall out."

In the dark, lanterns pumped up and down.

How many bodies were out there? Roy kept his eyes on the road. No time to count.

"They're not going to get us, damn it!"

Then he saw what appeared to be sawhorses, one on each side of the road, laid over with fence posts.

Bursts of light were closing in on both sides of the road. Small rocks were being pelted at the truck. Several hit the windshield. The back end of the truck slid slightly to the right.

"Hang on!" Roy didn't turn his head. "We're going through." He jabbed the accelerator, plowed into the barrier of old fence posts, heard the lumber crack and saw it fly.

Bits of gravel hit the truck body as tires bit into the roadbed and sent little stones over the road. Ahead several frightened men ran out of the way. The truck dove through.

Roy floored the Ford while Mike and Ole desperately held on, shoving the sliding milk cans back into place. He dared not let those cans tip and spill. Milk was needed next day at the dairy. He dodged chunks of lumber and stones.

"Roy," Mike called. "They're coming after us."

At top speed, Roy drove the panel truck to his home on Le Mars' northeast side, staying away from the dairy. On the phone he alerted his father and called police.

"There's been trouble with the Farm Holiday," Roy told police. "Come quick!"

While Roy talked with his father, the picketers drove up to the dairy.

"They aren't here," Fred told the men in answer to questions. As they spoke police pulled in.

"We have a report," the police chief looked squarely at the farmers, "a report that you men damaged a windshield." License numbers were jotted down. Disgruntled and grumbling, a dozen milling Farm Holiday members slowly moved away.

Next day the Globe Post ran a story about the incident, front page.

"Mike" and his mother, Miriam, pose outside the Harry "Mike" Wells home in the early 1940s

(Photo courtesy of Wells Family)

The Midnight Ride

Sixty four years later in June, 1996, Roy sat around the conference table with Fay, Fred Dale, Dan and Doug Wells at Wells' corporate office on Blue Bunny Drive. He related again, the oft-told tale. Four Wells' board members listened, interjected a comment now and then as the story progressed, and chuckled.

Each phase of the old familiar account drew mirth. Four men leaned forward to catch every nuance. With each detail hilarity grew. Clearly Roy was enjoying the telling.

"The midnight ride of the milk cans," said Fay. Five men exploded in another burst of jollity.

"And we had just gotten shatter-proof glass for the truck." Roy smiled, his eyes twinkling.

Chapter Sixteen

His whole purpose (as a writer) Civil War historian Bruce Catton once said, was "to re-examine (our) debt to the past"

Fay Wells

A War Intervenes

"If you boys have to go to war," Fred Jr. told his four sons in early 1940, "I'm going to sell the place." Youngest son, Fay, recalls his father's anxiety over losing sons and his business to war.

World War II needs had been decimating Wells' production staff. Personnel were leaving for war service, some inducted, some enlisted. The heart of Wells' Dairy was being threatened. Patriotism ran high.

Permitted by wartime government regulations to produce only the equivalent of previous years, Wells' Dairy business could not grow.

Sugar, essential to the ice cream business, became a rationed item. Troops got powdered milk since milk was rated indispensible for fighting men. Production of sweetened condensed milk had the go-ahead. These were not Wells' Dairy products.

Fay graduated from Le Mars High School in 1941, by fall had begun studies at Iowa State College, Ames. He elected dairy industry courses. The war in Europe ground on.

The menacing long reach of the U.S. Army grabbed Fay in the Spring of 1943. Drafted, he started basic training at Des Moines. Then he was sent to Fort Dodge and on to Georgia to train in chemical warfare.

"We were there a few months," Fay says, "then the Commanding Officer announced our U.S. Government had decided to abandon use of chemicals. So our unit was disbanded."

Thus a yoyo life with the Department of War began.

"I was ready to go," Roy said later. Roy, Harold and Mike were deferred since they were in an essential industry… food.

During high school years Fay dated but there "was not much time for socializing. Too much work at the dairy."

As a high school senior he became friends with Lucille Bray. Lucille had a job at the dairy in summer, bagging ice cream bars.

"We became close," Fay says. That fall he left for college, and "romance bloomed."

Lucille took a train to Georgia where she and Fay were wed in military-style. A few weeks later he was shipped to England.

In England he was interviewed "for background, skills, knowledge." They didn't need a dairy worker, he learned.

A good typist, his expertise was needed at the replacement unit. "First as clerk, later as classification specialist."

More change was in store. "We were shipped to France

in less than a year… to La Bourget. I was reassigned to 'casuals' as they came over."

Still more change. Sent into the Air Force as a staff sergeant, Fay says, "we went to Paris and Verdun… the year of the Battle of the Bulge… the German's last big push."

After the war, another move to a replacement depot. "We'd line them up for shipment… to a French seaport… to board ship and go back home, depending on the kind of work, the danger, how long overseas."

After a three-year absence, Fay, too, boarded a ship back to the U.S.A.

"I spent Christmas in camp," Fay says, "just before discharge."

While he was away a Wells' Dairy operation had expanded to Sioux City. Mustered out, Fay at times drove a transport truck for Wells' between Le Mars and Sioux City's new branch. He worked for a year as an ice cream salesman under Keith Wilcoxen.

A year later he returned to college at University of Iowa, Iowa City. He chose business courses.

Here Fay and Lucille's first child, Gary, was born. The little family lived along the river in a rented trailer. "It sure was cold in winter."

His G. I. Bill of Rights provided $90 a month for books and tuition, and earned for Fay a bachelor of science degree in commerce.

It's now 1949 and Fay is ready for the rigors of a steadily growing big business at Wells' Dairy. In Le Mars,

A War Intervenes

Patricia, Daniel and Douglas were born to the couple. A Le Mars apartment was their home.

At work, Fay became the dairy's first full-time salesperson.

Roy was general manager. "He gave me an old 1939 Plymouth to drive around and get new accounts. I suggested to Roy that we go to Maurice. We started with $20 in sales the first day. We went to Orange City (they had a strong local creamery), and Alton. I started the first out-of-town milk route.

"Doing most of the pioneer sales work, I developed accounts north up into Rock Rapids, Sheldon and Worthington, Minnesota."

Entrepreneurial in nature, Fay "started buying small dairies within 100 miles.... first Grubb Dairy in Kingsley, then in nearly every town, especially county seat towns that always had a dairy and a processing plant.

"Local dairies were not always happy about this." It was Fay who made the deals in "buying 40 dairies, most very small, the biggest in Cherokee… Alton, Marcus, Oyens, Remsen Bunker's Dairy, Granville, Struble, Brunsville, Rock Rapids, up into Minnesota."

By the 1960s Wells' had bought out nearly all small dairies in Northwest Iowa.

"Back then, most dairies had still been delivering house-to-house," Fay tells. "But grocery stores began to sell milk. They'd buy from us and sell for five cents cheaper. We couldn't sell it and deliver house-to-house as cheap.

"Volume was important. We could process for less money with volume. And more and more equipment was becoming available for high-speed processing."

It was typical of Wells' Dairy to introduce new methods and untried ideas.

"We were first to use the round milk cartons; they looked like a megaphone with a metal ring around the top. Quite expensive. But the cartons didn't have keeping qualities."

Wells' kept adding new products. Following trial and error the dairy "came up with a very fine cottage cheese," says Fay. "We built a reputation for cottage cheese."

And Wells' learned to "culture a good sour cream." Their biggest competitor, "Roberts Dairy made sour cream before we did. So we had Professor Neilsen come from Iowa State to help us.

"We were making the best ice cream that anyone knew of."

As the years flew by, Wells' reputation for high quality products kept growing.

Eventually drawn away from sales work, Fay became a vital part of production at the Le Mars plant.

During this time of rapid change, Fay, on December 3, 1968, married Betty Schipper.

Fay's sons, Gary, Daniel and Douglas, were graduating from college and becoming active in the business. And they were marrying. Ten grandchildren would eventually inherit the Wells surname. Two are part of the business so far.

On July 1, 1977, the company incorporated. Fay became president. "Before we didn't pay much attention to titles… it was all done by joint decision." They all worked

A War Intervenes

together as partners, Fay tells, without actually being a partnership operation. Now they were obliged to name their jobs.

Chairman of the Board and Chief Executive Officer became Fay's new title in the 1990s when Fred D. became President and Chief Operating Officer. "But," he points out, "it didn't change our jobs any."

Wells' Dairy was not sold, unlike the dairy operation of Fred Wells, Sr. in Chicago. There the first sale in 1905 to John Cunningham provided cash for homesteading in South Dakota. When the Chicago dairy was sold for a second time in 1929 to Bordens, it brought $5 million.

What advice does Fay have for those coming on in the business world?

"Dedicate yourself to education," he counsels, "and hard work… if you want to succeed. It's the same old advice.

"And family life is so important. Relationships are most satisfying. Making money is second. Real satisfaction is in companionship, personal responsibility."

The heritage left by Fred Jr., hard work, a good product, integrity, has been handed down to succeeding generations.

Fay's grandchildren are Benjamin J. and Andrew D. Wells, sons of Gary Wells; Melody R. Minthorn, Natalee J. Minthorn, and Huldah W. Minthorn, daughters of Patricia Wells Minthorn; Jacqueline D. and MacKenzie E. Wells, daughters of Daniel Wells; Kathryn E. Wells, Joseph F. Wells, and Sarah J. Wells, children of Douglas Wells.

Fay in uniform

(Photo courtesy of Fay Wells)

In the 1950s, the four sons that Miriam and Fred Wells once feared might be lost to World War II, were together for a family sitting in the Roy Wells' home: Harry Lee "Mike" *(left)*, **Roy, Harold and Fay**

(Photo courtesy of Suzanne Wells Stocker)

A War Intervenes

Working foursome (left) Mike, Fred Wells, Jr., Fay and Roy

(Photo courtesy of Wells' Dairy)

Fay and wife, Betty Schipper Wells

(Photo courtesy of Fay Wells)

Chapter Seventeen

The long ago is a part of each one of us

Anonymous

Pictured are Freds, Roys, Harrys and Michaels in the Wells family

See list on page 170

Fred Hooker Wells, Jr.

Harold Wells

Harry Lee "Mike" Wells

Fred Dale Wells

Roy Frederick Wells

Gary Michael Wells

Fred Hooker Wells, Sr.

Steven Roy Wells

Michael Todd Wells

Michael Cole Wells

Harry Cole Wells

Ronald Harold Wells

Joseph Frederick Wells

Fred Peter Wells

David Michael Wells

Michael J. Wells

Fred D and the Other Freds

Fred Dale Wells, grandson of Fred Hooker Wells, Sr., son of Harry C., and nephew of Fred Wells, Jr., came from Sioux City on weekends as a kid, with his parents and two sisters, Wilma and Marvel. They gathered with relatives at Aunt Nellie's in Le Mars.

Wells family members filled the home, spilled out from the kitchen to the porch and a backyard sometimes strung with lanterns to celebrate a special event… a birthday, a wedding, a birth.

Fred D. recalls his grandfather, white-haired and infirm, seated on a porch swing, while Grandma Clara and Aunt Nellie worked in the kitchen. (The grandparents lived with Aunt Nellie.) Sometimes Fred Sr. still played Santa Claus at Christmas gatherings.

The din coming from the Freds and Harrys, Harolds and Roys when the family got together raised the roof of the little house. Each visitor labored to be heard among contending voices.

"Fred?" someone might call out. The searcher could be seeking Fred D.; or cousin Fred Peter, Harold's son; Fred Hooker, Jr.; or Fred Hooker, Sr.

You might hear, "Harry, Harry?" above the social chatter. This could have been meant to rouse Fred D.'s father, Harry Cole; cousin Harry Lee, known as "Mike" to most; or perhaps have been directed to cousin Harold Raymond (Fred Jr's son.)

If Stella Wells Hampton (Fred Sr's daughter) were visiting from Washington State, it might mean her son, Harold.

And if Aunt Ella Mae (Stella's sister) had come from Wellsburg to visit her parents and called above the friendly racket for "Roy, Roy, answer me Roy," she might be asking for her son; her husband, Roy Hill; or her nephew Roy Frederick (Fred Jr's son.) Or perhaps Nellie's husband, Roy Chenhall.

Fred D. Wells, the family member who moved to Sioux City as an infant, was born at Doland, South Dakota. His father, Harry, had gone to Sioux City the year before to set up a new Wells' Dairy outlet.

His mother, Frieda, a Doland native, gave him birth on April 29, 1927. His sisters, Wilma and Marvel, also were born in Doland.

It was "tough going" for Fred's mother in Sioux City. One of thirteen children that had not known want, she found life in Iowa lonely and hard especially after her husband's death, June 22, 1943.

Young Fred started working at the dairy after school, loading trucks and getting behind the wheel as relief driver. He was learning the business.

In Sioux City, a manager was named after Harry's death, and Roy came from Le Mars to help.

As World War II began to crank down, Fred enlisted in the Navy. He served in the Pacific theatre first on a Navy destroyer. Then on a troop carrier.

Finally the war ended and Fred enrolled at Sioux City's Morningside College as a full-time student. At the same time he became a management assistant at the dairy, involved in sales and relief work.

Business at Wells' Blue Bunny (as it came to be known) demanded more and more of his attention. Fred finished two years of college and became a full-timer at Wells' Sioux City plant.

In early spring, 1951, April 1, Fred and Barbara Mulford of Kingsley were married at Kingsley's Methodist Church. It was a big wedding with bridesmaids and groomsmen. Roughly 200 attended. A reception at the church followed.

Their family began to build when Susan was born three years later on April 17. Michael arrived in May of 1959, and Gregory in October, 1966.

Meantime, Fred D. was exhibiting the Wells' ethic of hard work... learning the business from beginning on, and climbing the ladder in management and becoming, finally, President and Chief Operating Officer.

Fred Dale's grandchildren are Holly M. Sargeant, daughter of Susan Wells Sargeant; Michael J., Tiffany B., Rachel L., and Matthew C., children of Michael C. Wells.

Here are the Freds, Harrys, Roys, and Michaels in the Wells family:

- Fred Hooker Wells, Sr.
- Fred Hooker Wells, Jr.
- Fred Dale Wells
- Fred Peter Wells (Harold's son)
- Joseph Frederick Wells (son of Doug Wells, grandson of Fay Wells)

  Daniel Fred Renken (grandson of Roy Wells)

- Harold (Harry) Wells
- Ronald Harold Wells (Harold's son)

  Harold Hampton (Stella Wells Hampton's son)

- Harry Cole Wells
- Harry Lee "Mike" Wells

  Harry Fogarty (Wilma Wells Fogarty's son)

- Roy Frederick Wells (Fred Jr.'s son)
- Steven Roy Wells (son of Roy)

  Roy Chenhall (husband of Nellie Wells)

  Roy Hill (Ella Mae's husband)

  Roy Hill, Jr.

  Charles Roy Clemens (great-grandson of Roy F. Wells)

- Michael Cole Wells (son of Fred D.)
- Michael Todd Wells (son of David Michael)
- Michael J. Wells (son of Michael Cole)
- Gary Michael Wells (son of Fay)
- David Michael Wells (son of Harry Lee "Mike")

  Bradey Michael Wells (grandson of David Michael)

- Bullets indicate those pictured on page 166.

The Wells Spring

Harry Wells in 1940 prepared this float for the Labor Day Parade in Sioux City

In a serious moment at the Sioux City office, Roy Wells signs a document as a young Fred Dale Wells (left) and unidentified salesman watch the procedure

(Photos courtesy of Fred D. Wells)

Fred D and the Other Freds

Chapter Eighteen

Be economical: not parsimonious, nor stingy, but never go into debt

P.T. Barnum

Fred Wells Jr. holds Howdy Doody Bar, a Wells' product introduced about 1952

(Photo courtesy of Wells' Dairy)

A Frugal Life

"Have we got the money?" Fred Wells turned to Roy.

His son had just asked an important question. Roy had asked the same question before. Could they buy some new equipment for Wells' Dairy "so we wouldn't have to work such long days?"

Up at 3 a.m. and home after 11 p.m. was exhausting.

"No." Roy answered honestly. "We don't. But we could borrow."

"Forget it," his father replied. As he had before.

Fred Wells philosophized that you didn't spend money you didn't have. "Dad was very, very conservative. A lifetime of struggle made him cautious. My parents had it very tough. It was survival."

In hindsight, Roy sees this as good. "We owed no money and we stood the Depression. Dad was right on that one. We always had a top priority… a solid credit rating."

When business improved and his sons were able to carry on during his absence, Fred and Miriam began to travel. Using trailers, the two toured from coast to coast, from Canada to Mexico. "He owned thirteen different house trailers (over the years)." In Florida they parked the trailer at Miami.

The elder Wells didn't like hotels. "Too expensive."

Miriam was not as fond of travel as her husband. But as a loyal wife, she "went along with it," Roy recalls. "She did the cooking and he the driving." They were gone four or more months each year.

Later they flew to England and visited the graves of deceased Ralston relatives. Since Miriam's parents had come from Scotland, this was important to her.

"They'd fly to Europe or Hawaii and come home by boat. Dad would call from different places to ask how the business was going. 'It's fine,' I'd be happy to tell him."

During one trip, the two were on their way home from London on the Queen Mary, when Roy realized the day, September 28, was his mother's birthday.

"I phoned. Dad crawled out of bed and went to the radio room to answer. They had no phone in their stateroom."

"How are things going?" Fred asked as usual.

"Wonderful" Roy told him.

"It better be. For what this is costing you."

"Mother was pleased," Roy says. The bill for the call came for $11.25.

The sons were deeply appreciative of the role Miriam had played in establishing the business.

"In the 1940s Mother wanted a fur coat, but she didn't get it. In the late 1940s Dad bought Mother a large diamond ring."

Her sons bought a fur coat for their mother after Fred Jr's death.

Fay's early memories of his father and mother were of them talking about business. When Fay came home from school for dinner at noon, he'd hear of... "a new account at Remsen, Fairmont's trying to get one of our accounts... (or) we picked up a Fairmont account."

Business was a constant family topic and tonic.

"I'd hear stories in the evenings... at meal time. We all had supper together. Dinner at noon was the heaviest meal.

"Father was a very nice, wonderful, kind man. He didn't complain," Fay says... "he worked hard, didn't ask for much."

But "there was no fishing with us... no ball playing. He was always busy. And he loved to smoke big Red Owl cigars."

There were family vacations that made up for everything. "Dad had a 1937 Buick. I went on a couple of vacations with them... to Yellowstone and the Black Hills. Had a wonderful time.

"Mother was the sweetest lady. I never heard her say anything bad about anyone. If you can't say something good, don't say anything at all... that was her philosophy.

"She was really a very pure and decent person. She did her own housework and cooking. Dad would help with dishes... he always did the dishes. I dried.

"He was very kind to Mother. On Monday Dad did the laundry with her in the basement and saw that I helped. He was considerate...."

Fay has vivid memories of those earlier times.

A Frugal Life

In the early 1950s Fred took his wife to Mayo Clinic in Rochester, Minnesota. "Dad thought he'd go through the clinic too." Roy recalls. "He had been to see a doctor in Sioux City… he wasn't feeling well."

Fay was there when the X-rays came back from Mayo's. "Doctors decided to operate and remove part of his lungs… to stop cancer. Do an exploratory. But it was too late. The cancer was widespread.

"They couldn't do anything. They just sewed him back up.

"He came home and lived as best he could. But he was in pain.

"He died November 1, 1954.

"Mother was devastated when Dad died. She became ill. She missed Dad so much…. She was terribly lonesome… so lonesome.

"She lay in Le Mars hospital for a long time. She had blood and bone cancer. Doctors were never sure."

Fay, Harold, Roy and Mike visited her regularly. "I used to go up to the hospital each night," Fay tells. "I'd take her for a drive."

Just two years after her husband's death Miriam died. The date… October 27, 1956.

The two lie side-by-side in Le Mars Cemetery.

Fay speaks slowly and softly, "They loved each other very, very much…."

Fred Well's Funeral Today

Funeral services for Fred Hooker Wells, 69, who passed away Monday, November 1, were held today, Thursday, November 4, at 2 p. m. at the Le Mars Presbyterian church. Burial was in the Le Mars city cemetery, under direction of the Mauer funeral home.

Mr. Wells was a prominent Le Mars resident since 1913. He founded the Wells Dairy here and was a co-founder of Wells Blue Bunny Ice Cream Co. of Sioux City.

One of Mr. Wells main avocations was traveling. Places visited by Mr. and Mrs. Wells included Europe, Alaska, the Hawaiian islands, the Caribbean area. Mexico and Canada, visited in recent years since his semi-retirement.

Survivors include the widow, four sons, a brother, and five sisters. Also surviving are 14 grandchildren, and one great-grandchild.

Mr. Wells got his start in Le Mars in 1913 by buying milk and other dairy products from farmers, and selling it to customers who brought their own containers to his small place of business at about the same location where the big Wells plant operates today. The containers were filled out of large cans and the customers took their milk home.

Later Mr. Wells added a small delivery wagon, horse-drawn and made deliveries himself, as well as handling the processing end of the business.

From the very beginning, Mr. Wells insisted on absolute cleanliness from the dairy farm on through all steps to the point where the customer got the dairy product. Although Grade A milk was almost unheard of in those early days Mr. Wells' customers got as a matter of course, milk that would in all essentials, have qualified as grade A milk by the standards that were to be set in later years.

Mr. Wells introduced the glass milk bottle into use in Le Mars, and many years latter the paper carton, now so popular in foodstores everywhere.

He went down as when a lordly cedar...
goes down with a great shout... and leaves
a lonesome place against the sky

Edwin Markham

A Frugal Life

Fred Jr. and Miriam had spent three months short of fifty years together when Fred died November 1, 1954. When his own father died, one famous man is supposed to have said that a great tree had fallen, but that after a great tree is downed, the young struggling trees nearby begin to grow madly

Gathered at Roy and Kitty's home in 1950 were: (front row, left) **David** *(Mike's son);* **Madonna** *(Harold's daughter);* **Steven** *(Roy's son);* **Gary** *(Fay's son)* and **Bob** *(Mike's son)*
Row two: Nellie Wells Chenball; Frances *(Mike's wife);* **Miriam; Suzanne** *(Harold's daughter);* **Lucille** *(Fay's wife);* **Shirlee and Darleen** *(Roy's daughters);* **Eunice "Kitty"** *(Roy's wife);* **Betty** *(Harold's wife) holding their daughter* **Rita**
Back Row: Roy Chenball *(Nellie's husband);* **Ronald** *(Harold's son);* **Mike, Harold, Fred Jr; Fred** *(Harold's son);* **Fay, Roy and mother of Frances Marx Wells**

(Photo courtesy of Wells Family)

A Frugal Life

Mrs. Fred Wells Dies at Hospital; Rites Monday

Mrs. Fred H. Wells, 69, widow of the founder of Wells Dairy in Le Mars, died Saturday, Oct. 27, at Sacred Heart hospital. She was taken ill last April and had been hospitalized since.

Miriam Ralston was born Sept. 28, 1887, at Chicago. She moved to Wellsburg, S.D. in 1906, and her marriage to Fred H. Wells took place Feb. 6, 1907, at Chicago. The couple lived in Wellsburg until 1811, when they moved to a farm near Le Mars.

Mr. and Mrs. Wells took up residence in Le Mars in 1913, the same year Fred Wells began his milk business with a wagon and one horse. From these simple beginnings he developed one of Iowa's largest independently owned dairies and ice cream manufacturing plants.

Fred H. Wells died Nov. 1, 1954. The dairy he founded is now operated by his four sons and their cousin.

Surviving are the four sons, Harold, Harry and Fay, all of LeMars, and Roy of Sioux City; two brothers, Charles Ralston of New Brunswick, Canada, and Archie Ralston of Chadron, Neb., and 14 grandchildren.

Preceding Mrs. Wells in death, besides her husband were her parents, a brother and sister.

Funeral services were held at 2 p.m. Monday at the First Presbyterian church. Rev. James D. Ransom officiated. Burial was in the city cemetery under the direction of Mauer's funeral home.

SENTINEL., LeMars. Ia
WEDNES., OCT. 31, 1956

Part Three

Chapter Nineteen

I slept and dreamt that life was joy; I woke and saw that life was duty; I acted and behold duty was joy

Rabinranath Tagore

The New Generation is represented by (top, left to right), *Daniel Wells, David, Douglas, Gary, Mike and Greg Wells. Fred Dale,* (front left), *and Fay are second generation*

(Photo courtesy of *Dairy Foods Magazine*, photographer Bill Nellans)

A New Generation

Gary, grandson of Fred Jr., recalls when in 1954 his grandfather died, "I went to the funeral with Fay, my father. I was only six years old. We drove in a new 1955 white and black Buick. First time I'd seen that car. Pretty nice."

Gary carries this vague memory of Fred Jr. as one of the strongest he has of his grandfather. Above the physical is the powerful force of Fred Jr's work ethic. Work at the dairy was the heritage of Wells descendants. Gary was caught up in it at an early age.

As a youth he was already helping his Uncle Harold load a truck with "twin pops."

"Harold used to throw popsicles at me... twin pops.... He'd throw faster than I could throw them in the truck." Harold, too, is no longer alive, but he left his own forceful, amiable and amusing memory with Gary. Always anything connected with the dairy became pleasant and "fun," for Wells children.

Young Wells males grew up helping wherever help was

needed and hearing about the business from ground zero.

In the original plant at downtown Le Mars, Gary toiled at the milk cooler and with loading the ice cream truck.

During his first full summer at the dairy, he learned caseroom detail… here milk bottles were washed and placed in cases ready for reuse.

Gary was growing up. After he finished classes at Gehlen Catholic grade school his mother, Lucille, felt she wanted to place him in a military setting.

Gary was behaving in ways that concerned her… he was hanging out at Mullally's Pool Hall. The pool hall beckoned each day on the way home from school. "I got to be pretty good," he says, "at snooker tables."

His friend, Norm Cockrell, was Gary's "best buddy." Norm's father worked in the plant cooler. Naturally those two boys made a second stop after school… at the dairy to grab an ice cream cone or two and then to play hide-and-seek among boxes in the warehouse.

Gary spent a summer laboring in the plant cooler with the elder Cockrell. Then he packed up to attend St. Thomas Military School in St. Paul, Minnesota.

One hundred boarding students, some from foreign countries, were among his new acquaintances. Enrolled in the ROTC program, the new student from Le Mars found St. Thomas a "highly disciplined" seat of learning that had drawn many bright scholars.

"I hated the first couple of years there," Gary admits.

At St. Thomas he learned more than from books or

instructors. A new friend, another enrollee, Steve from Chicago, "taught me about smoking. He smoked Camels. He started me. We were 16. But I quit smoking twenty years later."

With Rick from Minnesota, he'd "go downtown to movies and drive the 'circuit.' I grew to like being in a big city on my own."

The Mississippi River was two blocks from campus. "There were caves and high banks, and parties there on weekends. One night we scaled the bridge over the river. Not smart but we made it."

Working during summers in Le Mars, he developed a "passion" for the dairy business. Following his stint in the caseroom, he moved to the ice cream plant making "pushups… tubes of round paper with sherbet inside, on a three-inch stick."

Summers were full of fun as well as hard work. At Lake Okoboji where the family had a condo and later a summer home, Gary drove a milk route. "I ran a combined retail and commercial delivery route to homes, restaurants and camps."

Uncle Roy had a trailer there and a big lake cruiser. Roy's son, Steve, Gary's cousin, had a job at Wilson's Boat Works. "We lived in Roy's trailer, went water skiing and had fun."

Schooling finished at St. Thomas', Gary studied at Marquette University in Milwaukee graduating in 1970 with a degree in marketing and finance.

"I was graduated one week and got married the next week, May 29, to Barbara Pick from Remsen. I knew her after I graduated from high school and we dated for four years while

I was in college."

Now a full-timer at Wells' Dairy, Gary says, "Fay put me under Bob Michael, our controller and cost accountant." Michael's job was to determine the cost of producing Wells' products. "He brought in a lot of new accounting methods. Michael had industry experience. He had been employed earlier at Borden's."

Michael's advice to Gary's father "was that if Fay wanted me to learn the business I should be Michael's assistant." This advice paid off. "Cost accounting gave me insight into marketing," Gary notes. "It gave me experience in various departments. Michael was a mentor for me. We talked about the dairy industry and the future of Wells' Dairy.

"I got a good grasp of what it took to produce products. It allowed me to strategize. How Wells' could grow… that there was not a lot of opportunity unless we expanded.

"We became one of the first to do private labeling for grocery chains and retailers, and to combine (and purchase other) dairies. We gained a lot of new business and that gave us a position in the market."

Nevertheless there were problems to deal with. "Our not being able to haul milk for more than 200 miles was a limitation to growth. But ice cream was different. We could haul it much farther.

"We sold to wholesalers in Iowa, Nebraska, Kansas, Wisconsin and South Dakota. This gave wholesalers an opportunity to enlarge their frozen food business.

"My brother Dan developed relationships with our co-pack customers. We began to produce frozen ice cream products for Disneyland (Mickey Mouse Bars), Sea World (Shamu Bars), Dairy Queen (Dilly Bars.) We helped large companies develop their product."

Gary's passion for the dairy business and marketing grew. As Information Systems Manager (earlier known as data processing manager) he was in charge of a new IBM system and computer department.

His enterprising nature unfolded and took off during the next five years. "I wanted to sell ice cream programs." As a consequence he went into sales, developing warehouse distribution… "giving us more growth."

Next step was to Vice President of Marketing. "We developed new novelty products as well as a premium ice cream program. We then developed a sales force to present the program to new and existing customers."

From Fred Jr. had come, handed down to each generation, emphasis on quality of product and a goal to make the product a better value for the money than other competitors offered.

"We put in state-of-the-art production lines to keep cost low. The key" Gary says, "to producing quality products."

In 1989 Gary moved to Executive Vice President and Chief Marketing Officer. Wells' Dairy introduced specialized product programs for Dairy Queen, Olive Garden, Red Lobster, Applebee's. A street vending product line for large cities was begun.

"Part of our success," Gary says, "is a work ethic that hires good people who stay and are involved… superior people.

A New Generation

And for me also the support of my wife and family."

Through eminence in marketing and a top-of-the-line product, Wells' has grown to encompass business in 48 states and 30 small and large countries world-wide… countries that do not have the know-how or the "infrastructure," he points out, to produce and market a fine product.

Even with heavy responsibility, Gary found time on weekends at Lake Okoboji, for family, wife Barbara; Benjamin, a graduate of University of Southern California and a Wells' sales representative in Phoenix, married to Anne; and Andrew a student at Arizona State University.

There are roots in Remsen where Barbara, one of nine children, is the daughter of a veterinarian. Their families spend time together at Lake Okoboji where Gary and Barbara have a second home.

Fred Jr. would be proud to see where Wells' Dairy has catapulted… to one of the five largest ice cream producers world-wide. Even so, it might be difficult for him not to pause and ask questions about the need for expansion. He might shake his head and ask, "Do we have the money to do it."

Oldest member of the new generation, Gary is the first-born of Fay and Lucille, photographed here in the home of Gary's grandfather and grandmother, Fred Jr. and Miriam

(Photo courtesy of Shirlee Wells)

Gary Wells, Executive Vice President and Chief Marketing Officer

A New Generation

Parents of the groom, Gary and Barbara Wells (center) are photographed with their sons, Andrew (left), and Ben, the groom, on Ben's wedding day

The bride, formerly Anne Cummings, and Ben stand for the photographer. Anne holds her wedding bouquet on their day, September 6, 1998

Chapter Twenty

Youth... flashing like a star out of the twilight

Willa Cather

Dan and his family live in the country where they raise horses and dogs. Shown are (left to right) Dan, MacKenzie holding her cat, Lisa and Jacqui with her dog

(Photo courtesy of Dan Wells Family)

Keeper of the Capital

Dan Wells and his young friend Denny Lenihan, Jr. had an exhilarating after-school routine. Cousin Robert "Bob" Wells had sold Dan and Denny an old bronze-colored Ford "flathead eight-cylinder," somewhat mangled with age.

That "crate" had a unique riding style. "When you stepped on the gas the front seat would fly back," Dan says. "It would lunge… back and forth… it wasn't bolted to the frame.

"We'd take it to the site of our fluid milk plant on ten acres out of town." A pile of dirt there had a gradual slope.

"We'd head up the slope and fly off a cliff on the back side of the dirt pile. All the tires would blow out. We'd fill them up with air and do it again.

"Bob had a 22 pistol. He'd pick off gophers" while the old car swung along, back and forth.

"Joe Sutton was the police chief," Dan recalls. "He didn't come after us. He just took that old car away. We never saw it again. He confiscated it."

Cousin Bob was a prankster and practical joker. "We were working in the caseroom at the dairy," where they unloaded cases of milk, washed bottles and set up for production.

"Bob was a character. On a hot July day he came in with some orange juice. I grabbed it and drank. I thought it was kind of sour. About ten minutes later I was swinging in the wind. That was my first run-in with alcohol.

"And that was Bob." Cousin Bob later moved to Nevada where he still lives. Bob is the son of Harry Lee (Mike) and Frances Marx Wells.

Dan's growing up years were filled with fun and good friends. Coming home each day from Gehlen Catholic Grade School the boys took the alleys through Sixth, Fifth, Fourth and Third to the dairy on Second Avenue.

"We knew all the dogs along the way. And the lilac bushes. The alleys were full of lilac bushes.

"We'd go to the dairy to get a fresh supply of ice cream bars. As kids we were interested in ice cream. We'd steal rolls of bottle caps and throw them around."

In the dairy there were "giant candy bars of chocolate. We'd hijack those candy bars once in a while."

In making stick novelties, molds were filled with ice cream… lifted out and dipped into hot chocolate. Small freezers and a brine tank with salt and ice, and an ammonia process cooled the ice cream.

Novelties were put into open-ended bags. (Now, Dan points out, there is "a seal-wrapped process.")

Milk cartons at that time were bathed in large pans of

hot wax to seal, then placed into a cooler. "We'd stick our fingers into the hot wax and have waxed fingers. That was fun. We played a lot around the plant."

At 14, Dan walked into his father's office and asked, "Could I go to work here, Dad."

"You'll have to go to the (Plymouth County) Courthouse," Fay told him. "Get a work permit."

"That scared me. Three days later I told Dad I had gone to the courthouse. He looked at me as if he knew I hadn't. But he told me to go into the milk plant.

"I started washing half-pint bottles and helping with the school milk program. That program was a great boost for the industry and for milk farmers.

"I wasn't interested then in work," Dan confesses. "Just in being a part of the operation.

"Wells' had bought a semitruck that would bring back empty wire cases. They were heavy." It was there in the caseroom that Dan worked first with his cousin Bob.

After graduation from elementary school, Dan attended Maur Hill Boarding School at Atchison, Kansas, an old river town across from St. Joseph, Missouri. "You could do a lot there and not get caught. It was an all-boys' school three blocks from a girls' school." In Kansas Dan once met Bob Dole before Dole became nationally prominent.

After completing two years at Maur Hill, Dan told his mother that since Gary was going to Marquette University, he (Dan) would like to go to Milwaukee Dominican High School to be closer to his brother.

Keeper of the Capital

"It was first class, one of the friendliest experiences of my life. They were big in basketball (at Marquette). Al McGuire, famous as coach, really invented arguing with referees. He was aggressive. They had fabulous basketball teams during 1968, '69 and '70."

After two years at Dominican High and one year at Marquette, Dan went on to study at Fort Collins (Colorado), and the University of Iowa.

Back in Le Mars and retired from route jobs, he began his service at Wells' Dairy first as a buyer and in purchasing, slowly and steadily moving up to Contract Packaging and Procurement and to Senior Vice President and Treasurer.

In 1982 Dan met Lisa Athens, a Le Mars girl. Where else would small town young people meet, if not in school, through mutual friends and at a local restaurant?

In 1984 the two married after a year-long engagement. Five hundred guests gathered at her parent's home, the Roger Athens, on Central Avenue SE, for an outdoor wedding.

"It was a gorgeous day," a memorable June day… the twenty-third. Seven bridesmaids and seven groomsmen attended the couple.

"Roy came in his 30-year old Lincoln Continental, a collectible," that the two used that day. "I was proud Roy would be my best man," says Dan. "It was a great wedding."

Lisa and Dan now live on 80-acre Big Creek Ranch out of Le Mars, with daughters Jacqui, 12, and MacKenzie, 8, a barn full of Belgian horses and ten Tennessee Walkers; Rottweilers, St. Bernards and German Shorthairs. A creek runs through the property past their newly-built home. Seven

Emden's white geese and some ducks quack and hiss at newcomers.

Lisa, a graphic artist, has collaborated with Dan on designing frozen yogurt cartons, stick novelty packaging and T-shirts. Together they created the Mickey Mouse cookie that DisneyWorld uses. Lisa also paints in oil and draws.

The two are active in community affairs… Dan has been interested in and contributed his efforts to Plymouth County Historical Museum; he is a former Le Mars Chamber of Commerce board member. Lovers of the environment and of preservation they are members of Pheasants Forever. Lisa has been involved in Floyd Valley Hospital Auxiliary. Jacqui is a 4-H Club member.

"We travel a lot," says Lisa. Dan, speaking for their rural life, feels "It's ideal for the kids."

A beautiful dog, a beautiful horse and a beautiful day in the country with Dan Wells

(Photo courtesy of Dan and Lisa Wells)

Dan and Lisa Athens were married in 1984

Senior Vice President and Treasurer, Dan Wells

A bobsled full of Wells' Dairy employees are given a snowy Christmas treat through the Wells' corporate office parking lot. Duke and Dan, Belgians belonging to Dan and Lisa Wells, pull the sled. Sonny Athens holds the reins. Note Wells' semitrucks in the background

(*Le Mars Daily Sentinel photo by Judy Bowman*)

Chapter Twenty One

Each generation owes the next knowledge of the past. It's a shared bond from which we gain strength to face tomorrow

David Buscaglio

The Doug Wells family, all smiles, are (left to right), Sarah, Ellen, Doug, Joe and Kathryn

Fire, Acid And Work Fulfillment

Doug Wells, at age three, was too young to put out the blaze. On that hot day in August, all the Wells relatives had gathered in the backyard at his parents' home for a birthday party when the tablecloth caught fire.

Candles were being lighted just as unexpected breezes caught the birthday napkins, stacked ready for picnickers, and aimed them straight at candles on the birthday cake.

Suddenly the table was ablaze.

Brother Gary and sister Pat were standing close by. "It was fun," Doug says now. Too young to have heard much about the disastrous Fourth of July fires in 1936 that set small Oyens blazing and decimated Remsen's business district, he jumped with glee as he watched adults scramble. He knew little of adult anxieties about fire during sizzling summer months.

"Mother put the fire out." he says some forty years later.

Those were great times for family gatherings and family

vacations. "Mother's brother and sisters lived in California. Rosemary and Frank Bray, mom's parents, and her sister Delores, lived in New York City. But Rose Marie, Rita, Lorraine and Joe lived out west."

Doug became a well-traveled youngster. Family trips took them to Yellowstone, New York and Los Angeles and to Tacoma, Washington to visit relatives. "They were a lot of fun. Some of my fondest childhood memories are of these long expeditions that pulled the family closer together… quality time with each other."

As a youngster Doug, like his brothers, sometimes went into the ice cream plant on weekends with Fay, "checking things."

"Don't stick your hands in there, Doug," his father told him on one plant tour when Doug was seven.

They were passing an acid-sanitizer in a stainless tank. The liquid was crystal clear and had a slightly blue tint that attracted the small boy.

"I stuck my hand in the acid. I was whimpering when Dad saw me shaking my hand. He washed it with soap and water. He was mad at me."

Grandfather Wells, (Fred Jr.) died when Doug was barely two. He has a vague recollection of him.

"When I was very young I have this memory of a big man walking toward me. It must have been Fred, Jr. We were on the back porch of a house next to the plant, the house that was relocated. Every once in a while Fay still drives by and looks at the house and remembers and talks about it."

There are fond memories for Doug, of his Uncle Harold. "He and Betty lived in a little house at Cleveland Park. We'd go up on Sunday for coffee and play around his house." Harold had responsibility for the Wells' truck fleet. Harold died suddenly one Sunday while "checking things" at the plant. He was 63.

Other role models helped to keep the Wells youngsters on the straight and narrow. Uncle Mike (Harry Lee) was one.

During high school Doug worked after hours in the ice cream plant. "My buddies and I were going fishing. We took an 'extra' Wells' Dairy van for a short trip to the fishing hole."

Bringing the van back, "we pulled into the driveway where Mike was waiting. He had responsibility for maintenance of the Wells' trucks. Boy, did I get chewed out for 'personal' use of a company vehicle ."

In the production line, during summer at the milk plant, Doug worked with Uncle Roy. "We'd set up equipment, run ice cream sandwiches for seven, eight hours and then clean up .

"Roy would tell me how important it was... that everything be absolutely sterile... and how important it was to have small ice crystals and small, fine air cells. The smaller and the more numerous the air cells are, the smoother the body and texture of ice cream. Our primary objective was good smooth texture."

Doug developed a lasting relationship with Roy. "He had a lot of patience. We'd walk through the plant. If he found an employee not paying enough attention, he'd point it out about the air cells… if the cells were too big. I enjoyed conversations with Roy, and the work."

After graduation in 1972 from Gehlen Catholic High, Doug attended Le Mars' Westmar College and continued to work at the plant in summer. "It was tough getting used to the hours… 2 a.m. to noon. Once I woke up at 2 p.m., got up and dressed for work. Then I heard the birds chirping…." He chuckles. "Outside, birds were chirping in bright sunlight. I knew it wasn't 2 a.m."

Later at Creighton University, Omaha, he studied for degrees in Business Administration. At Creighton a Le Mars girl, Ellen Ahlers, whom Doug had known at elementary school and Gehlen High days, had come to study.

"In 1973 at Creighton we met at a party. I got interested, I took her out and kept dating her through college. We had so much fun."

After Creighton, dairy courses at South Dakota State at Brookings sent Doug and Ellen in opposite directions. She had taken a job at Spencer, Iowa teaching special education. But he pursued her.

After driving miles between Brookings and Spencer to spend time with Ellen they married in the summer of 1978. Ellen transferred to Le Mars Community Schools where she taught for five years.

In 1983 Wells' Dairy purchased a plant in Omaha changing their lives again. The new property was renovated and remodeled and prepared for processing milk, fruit juice, and yogurt. Doug commuted to Omaha.

At an international Ice Cream Association meeting in Washington, D.C. a proud father introduces an eager Kathryn to her first Wells Blue Bunny Ice Cream cone

"Long hours… 16-hour days… home on weekends." The two bought a home in Omaha and moved there. Kathryn, their firstborn, arrived June 16, 1984.

In Le Mars meanwhile, the north ice cream plant on Second Avenue NE cried for expansion. A major new ice cream structure unfolded with David Wells, Mike's son, and Fay carrying the load of planning.

In 1986 a new mix department, ten more production lines, and a high-rise freezer doubled the plant's capacity on Second Street SE.

Doug was needed in Le Mars. He had moved to plant manager in Omaha. He and Ellen now came back to Le Mars and lived in a little house near the north plant, where Joe was born October 29, 1986. And Sarah on August 14, 1988.

"Tight and expensive housing" inspired the couple to

Fire, Acid and Work Fulfillment

look at building their own home. They purchased a lot in an expanding development and moved to a new home on Joe's second birthday.

Doug's time with Wells' had now reached eleven years since he began full-time work. Production, sales, special order deliveries and setting up promotions, had been his prime concern.

One of his responsibilities became to keep Wells' in compliance with governmental regulatory procedures and requirements. Formal quality control was a new concept at Wells' Dairy. Standardized ingredient quality, microbiology and numerous product tests at the plants were his expertise.

Gradually he spent more time at the ice cream plant than at the milk plant. "As we grew it became obvious we needed a formal quality control department." Doug became quality control supervisor of the Le Mars and Omaha operations.

Wells' had been growing rapidly. Now "we were running Gold Rush bars under the Betty Crocker label for General Mills and delving into ice cream novelties." Glacier 1500 was innovative technology for that time.

In 1990 Doug, promoted to Vice President of Manufacturing, worked with Fay and Dave in planning the new south ice cream plant.

Finding an appropriate location for a 165,000 square foot plant was not easy. But the first shovel of dirt was spaded and a year later, July 2, 1992, the first ice cream bars came off the Glacier Omni 3000 ice cream machine. Four production lines were increased to eleven in 1993. Five more were added by 1994 and five more in 1995. Twenty one lines moved daily at the south plant.

Doug became Senior Vice President of Manufacturing.

He is generous with his credit to personnel who have helped growth at Wells' Dairy, Inc. "They have contributed significantly to the company's success. They work in concert with veteran ice cream makers and that has produced a great combination.

"We're working on improving the 21 lines." Doug's enthusiasm is endless. He projects this with a wide smile.

Need for more storage to build inventory of ice cream products arose and construction began to house a 2.8 million cubic feet ice cream storage system. The ASRS (Automated Storage Retrieval System) stores 10,300 pallets of ice cream at -20 degrees F° and uses four unmanned, computerized cranes to move 250 pallets an hour… in and out.

"Technology, innovation and unpredictability of the future, drive change today at Wells' Dairy," says Doug. Today's industry demands talented people to formulate delicious tasting "fat free" and "no sugar added" frozen dairy desserts. "Proprietary processing techniques result in flavorful homemade vanilla and homemade chocolate, distinctive and unique premium ice creams.

"Driving change and continually improving our resources and capabilities, strengthening Wells' Dairy and making it more self-reliant is hard work and great fun.

"Improving quality and productivity is rewarding but developing professional technical and engineering resources is a real challenge," says Doug. "It is one we will continue to develop as rapidly as we can.

"State-of-the-art freezing technology," is used now since two new automated high speed novelty production lines were added to the South Ice Cream Plant in early 1999.

"Six new production lines" will enhance processing and production capacity by the end of 1999."

Doug Wells' mind still works seven days a week and every waking hour in an industry that keeps striding, galloping forward and onward.

Dr. Ed Oamen, (left) Dr. Verner Nielsen, Dairy Science department head at Iowa State University and Doug Wells, Senior Vice President of Manufacturing, pose following an important meeting

Doug and Ellen, (right) pose with customers after a golf outing in Sydney, Australia. Golf is Doug's hobby

Fire, Acid and Work Fulfillment

Chapter Twenty Two

The passing moment is all that we can be sure of; it is only common sense to extract its utmost value…; the future will one day be the present and will seem as unimportant as the present does now

W. Somerset Maugham

Wed to Marilyn Moran April 30, 1988, David posed with Marilyn's parents Donna and Gene Moran (left) and his parents Mike and Frances Wells

Wells On The Fast Track

In 1905 when Fred Wells Sr. and Clara, his wife, prepared to homestead in South Dakota, journeying slowly and ponderously from Chicago by train, horse and wagon… at the same time at Kitty Hawk, North Carolina the Wright Brothers in their air machine, flew 24 miles in 35 minutes. Less than a mile a minute.

Only a few years later in 1911, Roy Harroun moved 74.50 miles per hour in the first Indianapolis 500.

Now jump to 1996. One weekend in 1996 David Michael Wells sped 248 mph in the cockpit of his drag racer Plain Vanilla… a Wells' Blue Bunny Ice Cream logo branded on its rear flank. Wells did this, helmeted and hunched in his vehicle, within five seconds from a standstill… the first Wells to travel so fast on the ground.

Weekdays David is at Wells' South Ice Cream plant in Le Mars, overseeing an enormous windowless building, 356 by 100 feet, that is really one large freezer. Here frozen ice cream products are produced and on-the-ready to be loaded at an adjoining dock. One of 767 trucks crammed with ice cream products leaves the dock every 13 minutes. Wells' trucks, some 200 of them head to points all over the USA.

David, a third generation Wells, grandson of Fred Jr. and Miriam, has come along in the same tradition of family, work and play that began as a youngster who visited grandparents every Sunday afternoon.

His grandparents lived, he recalls, "in a house next to the (downtown) north plant where we'd go to see what they were doing. I remember him (grandfather) as a large man who smoked a lot of cigars."

David made the regular Sunday visits with his parents, Harry "Mike" and Frances Wells and brother, Bob. Visiting with family was paramount on Sundays, holidays and family occasions. "We got together at Christmas, usually at Roy's house in Sioux City."

David had a strong inclination for the mechanical. At 14 he applied for a work permit and began as all young Wells males by mowing the yard. His bent for the mechanical made it easy for him to enjoy covering pipe insulation. He painted cabinets in the plant.

"At 15 I started on the production line at the north plant," David says, "ran the bottle washer at the milk plant, worked in the cheese department, packaging... on stickless novelty lines."

While honing a genius for the mechanical David "worked for a time at a body shop in Le Mars, for Johnny

Mayrose… and Hank Livermore Construction."

After graduation from Gehlen High School, college was a must. The years 1967 and 1968 were spent at Westmar College.

"Then I was back on the novelty lines at the north plant, on the Gram machine with Ron Delperdang. We put ice cream bars on a stick. In 1970 a second Gram machine was installed at the downtown plant. Ron went to that machine."

Soon need for a second shift emerged. "Noon to 10 p.m. In the summer of 1972 I worked second shift. Worked that for a year or two. Ron became a supervisor."

College days over, "I went into maintenance at the north plant," David says… 1973-1974. "I'm still there. (as well as at the south plant.) But there were just two of us then, Dick Kommes and me. Next couple of years we worked double shifts."

Responsibility came early and accelerated on a fast track. "If they had a problem they'd call me. I'd get the night calls. Next I became head of maintenance and was doing floor layouts, installing new equipment, figuring out refrigeration requirements."

By 1984 David notes "major expansions were needed at the north plant. Wells' Dairy bought homes on the west side, the north side, the east side of the old plant.

"We had to tear down five homes, doubled our capacity, built an engine room for refrigeration and another freezer." The mix department was enlarged.

The north plant is now one large building… 109,000 feet ground floor and 44,000, second floor. "Takes up one city block."

In 1991 the South Ice Cream Plant (south of Twelfth

Street) was completed. "A duplication of the north plant, but larger, faster." A new tradition of need for speed had emerged. Competition pressed, bearing unusual weight. "We needed more capacity.

"Basically we do the same thing at both plants, except for a few items." David was sent to handle controls of the new location that sits on 112 acres. The big white refrigeration building reaches 98 feet up into the ozone and can be seen for miles when you drive into Le Mars from the west.

Two low-bay freezers, a materials handling room, and cold storage space... 227,000 feet of it hug the big building.

An Engineering and Research Center on the new south acreage followed in 1996. Employees under supervision of David Wells (Bob Book, Tony Chihak, Dick Kommes, Curt Staab, Mark Jordan) create blue print drawings of new production lines. A Computer Aided Drawing System (CAD) under fabrication supervisor Bob Book hums while personnel keep abreast of changes required for each upgrade of Wells' machinery.

Machinists keep dolly repair moving, conveyers conveying and machines maintained.

An automated truck wash on the big acreage scrubs 200 mega-sized semi-trailer trucks ablaze with sparkling Blue Bunny logos. Every ten minutes one sanitized truck, loaded with ice cream products leaves the loading dock for destinations all over the USA.

At Wells' "We try," says David formerly Ice Cream Plant Manager, now at the controls as Vice President of Engineering, "we try to make what is needed to produce the product that people want."

David lives north of Le Mars with his wife, Marilyn Moran Wells. He has three sons, Michael, 30, married to Kristin Kaiserlik, Brian, 27, married to Wendy Clark, and Neal, 26, married to Karen Knuckles. He has two stepsons, Marc Alan Scott and Erik Stephen Scott.

When the rigors and the stresses of his responsibilities lay heavily, David escapes to Plain Vanilla and cruises along at national drag races held from Ohio to Oklahoma. Refreshed, he returns to what he knows best and excels in, where the big refrigerator speeds to meet demands of an accelerated society.

David has three grandsons, Bradey Michael, son of Mike Wells, Tyler David, son of Brian Wells and Caleb James, son of Neal Wells.

Attendants at his marriage to Marilyn Moran, April 30, 1988 were David's sons, (left), *Neal, Brian, (David) and Mike* (right)

Appropriately named Plain Vanilla, is David's 26-foot dragster. During his 17th year of racing, in June, 1996, he placed second at Memphis National races. Helmeted, David is seated in the cockpit

(Photo courtesy of Wells' Dairy)

Wells on the Fast Track

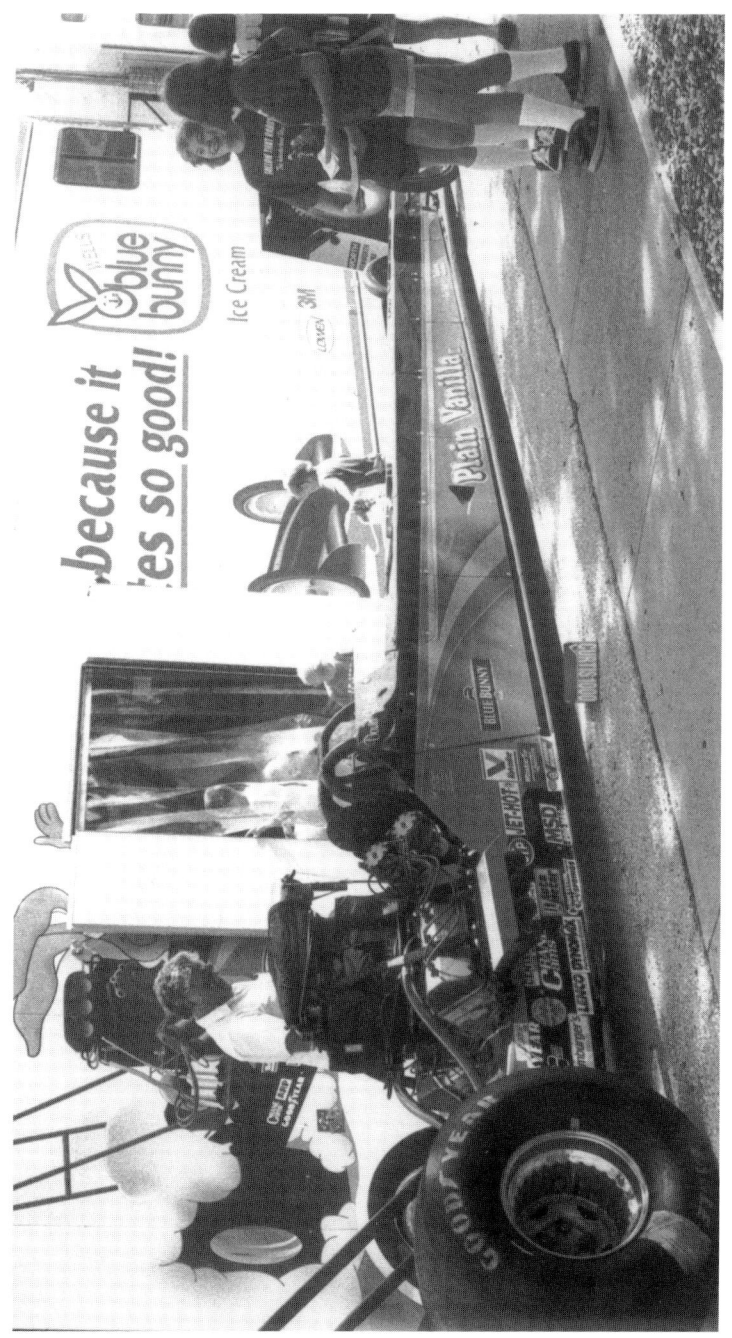

Plain Vanilla, dragster was on display during a reading program at Le Mars Public Library attended by 200 young readers. David Wells stands left with his racer that has hit 244 mph in a quarter mile in 5.67 seconds.

(Le Mars Daily Sentinel Photo)

Chapter Twenty Three

Business is very much like religion: it is founded on faith

William McFee

A third generation Wells, Mike poses with his family, (seated) son Michael, wife Cheryl, (standing) daughter Tiffany, Rachel and his youngest Matthew

(Photo courtesy of Mike Wells family)

Tough Times And Good Times

Michael, son of Fred Dale and grandson of Harry Cole Wells, was among the clan that gathered often at Aunt Nellie Wells' home in Le Mars.

Young Mike and his family toured up from Sioux City where Fred D. and his father before him were involved in the Wells' Sioux City venture. They brought with them his grandmother, Frieda (Harry's widow.)

Aunt Nellie no longer ran Nellie Wells' Beauty Shop. She had married Roy Chenhall and the two moved to a large three-story home (with an historic background) across from Andrew Carnegie Library on Central Avenue.

"We'd play up on the third floor of Nellie's home"… we being Michael's sister Susan, and brother Greg.

Like all Wells children they grew up closely connected to the growing extended family and the larger Wells' Dairy operation embracing the Sioux City and Omaha enterprises.

Fred D. headed the Sioux City operation. He and his family lived in Morningside.

A family tradition of children accompanying their father on business excursions is among Mike's privileged memories.

"Dad would take Susan and me along with him to the plant. We'd ride the elevator, big enough to carry large trucks up and down. We were fascinated with that building.

"Dad was responsible for Sioux City and South. He spent time in the Omaha market, developing accounts.

"He'd get a call late at night from International Garage to drive a repaired truck back to the plant. He was called out a lot on weekends. Local stores would run out of our products."

At age twelve Michael was mowing the acreage around the Wells' Dairy Steuben Street branch. "I'd help load trucks, sweep floors. Dad had a 'work yourself up from the bottom' (credo). You needed to prove you could do the work."

When he reached high school he began to work inside the branch. Gradually Michael took on more and more responsibility.

Sensing where this seemed to be leading, Fred warned: "The business is demanding but rewarding. Make sure it's what you want to do for a lifetime."

While giving work their strongest and best, Wells family members always knew how to enjoy life outside toil.

"We'd boat on the Missouri River at Sioux City, water skied. Ran around in shorts. It was fun."

After graduation from East High School in 1977, Mike married Cheryl Abramo, his high school sweetheart, and began full-time at the Sioux City plant.

He drove a school milk route and made special deliveries… and enrolled at Morningside College. "Classes in the morning, a delivery route in the afternoon. Until 9 p.m. And on weekends." Fifteen hours a semester at school, 45 hours a week of work was his rite of passage into the family business. But fun work.

The couple lived in Morningside where their son, Michael, now 20, was born.

The dairy "was open Saturday and a half-day on Sunday. We packed meat and ice cream in dry ice for people. And ran special orders."

Out of college Michael and his family moved to Omaha. Cheryl was pregnant with their second child, Tiffany.

"I accepted a sales position… my first solo experience. Dean Alberts handed me a map and a set of keys. I was scared to death."

Given a big territory, Shenandoah and Red Oak, Iowa, Rockport, Missouri and Falls City, Nebraska Mike spent nights away from home. Or got home late.

"After I'd driven 2,000 miles in my new company car, the dealer advised me to bring it back in. In two weeks I had driven 2,000 miles.

"I made my share of mistakes, but," says Mike, "Dean showed me the error of my ways. He covered for me.

"There were three of us salespeople. Every third weekend I'd have to work. Those were wild weekends." Weekends that cut a swath in his memory.

"One July Fourth a big semi hauling our milk hit the ditch outside Council Bluffs. The driver had fallen asleep. It was 5 a.m. when I got the call.

"I had never driven a semi before but learned how that day. We cleaned up after it… milk was everywhere. It was 102 in the shade. What a mess. Omaha had no milk that day. We got a truckload from Le Mars and delivered it that night."

Sometimes unappealing jobs come along even in an industry that you love.

Writing orders for the cooler didn't inspire the salesman. But that was his job for a time. "You'd inventory products and guess what you'd need, then write an order. It was hard to anticipate needs."

After 30 days he was told: "You're a better salesman than an order writer."

In 1984 an Omaha milk plant had been purchased. A fleet of trucks operated out of, and was maintained in, Omaha. Mike was in charge of the fleet.

Still in sales, he was now based closer to his home.

In 1985 the Michael Wells family moved back to Iowa… to Hinton, north of Sioux City. In Hinton Mike spent four years on City Council and served on the golf course board. "We got to know a lot of nice people in eight years." Their third child, Rachel, was born in Hinton.

While living in Hinton Mike moved up the sales ladder. In Le Mars he worked as assistant to Le Mars Sales Manager, Dick Gaul, and began to assume corporate responsibility for the transportation department.

"I worked with Fleet Manager, Wayne Schwartz," he added, "taking corporate responsibility for the transportation department and learning about corporate sales operations on a national scale."

Mike was promoted to Sales Director, "becoming more involved in Wells' ice cream business all around the country."

Today Mike's title is Vice President of Sales and Transportation.

"Every day there is something new (in the industry). I love this company and the work I do.

"Cheryl and I wanted to be in Le Mars. It's a neat community." Matthew, 3, was born after the move to Le Mars in 1994.

"Cheryl and I have raised our kids in a Christian home. We believe in the need for God's place in our life and our home; wherever we have been and we have placed our trust in Jesus Christ.

"Our faith has brought us through the tough times and helped us enjoy good times."

As a youngster, Mike was served a Blue Bunny birthday cake

During college years, Mike also worked for Wells' Blue Bunny

Three generations of the Fred Dale Wells family have loved the river. Here are Mike, holding Michael Jay, and Fred D.

The Wells Spring

Chapter Twenty Four

The winds and waves are always on the side of the ablest navigator

Edward Gibbon (1737-91)

Beloved grandmother Frieda Wells, wife of Harry Cole Wells, and mother of Fred Dale Wells, came to Sioux City with her husband, from Doland, South Dakota to bring about the Sioux City operation in the 1920s

(Photo courtesy of Wells family)

River And Spring

Gregory Arthur Wells was "born to the river" and into Wells' family traditions. Always there was a closeness that bound members together in fun, good humor, good will.

The youngest member of the clan is one of eleven involved in Wells' daily activities at 1 Blue Bunny Drive, or somewhere in the U.S.A.

Greg is no different from the others yet is no cookie-cutter replication. He is an individual with his own aura of dominance… who can be recognized at the same time as coming from similar roots.

The youngest son of Fred Dale and Barbara Lou Mulford Wells, Greg springs from the Fred Hooker Wells, Sr. branch that began in Wellsburg, South Dakota. His grandfather, Harry Cole Wells left Wellsburg for Doland, South Dakota about the time that Fred, Jr. and Miriam left for Iowa. All were seeking a better life.

Later Harry Cole, too, left South Dakota and came to Sioux City to open the Wells' operation there. He died when his son, Fred, was 16. "I never knew my grandfather." Greg regrets that.

Frieda, Harry's widow, lived near her son and his family in Sioux City. Freda and Greg her grandson bonded in a special way. His memories are vivid.

"She loved to walk and we walked with her. She loved mashed potatoes and gravy. So I became a big mashed potato lover. I got to know her better than most kids… I spent quality time with her.

"She lived in Morningside first. We'd go to visit her. She had an old lawn mower, a push mower that I'd push. That was fun.

"We used to kid her a lot. We'd see her walking to get groceries or go to the post office, always with her head high.

"But when she crossed an intersection, she'd walk with her head down." She'd plow through.

"Later she lived with us but in her own separate quarters, for quite a few years."

Young Greg and his brother, Mike, often accompanied their father on Wells' business trips. "I enjoyed spending time with my dad. When I was a youngster I went with him once a week. He'd make sales calls and I'd wait for him. It's special to remember."

Fred stayed close to his sons. Not a good student, Greg admits he was " more interested in playing out of doors. Studies came hard. I brought my report cards home and Fred would sit down with me… tell me the facts of life.

"We had good parents…. (There also was Susan, an older

sister.) They listened to both sides. I have the utmost respect for them. My love for boating and being on the Missouri River… I inherited from my parents." Greg was born to the river at Sioux City.

Wells menfolk had the abiding ability to make work appear as fun… hard but fun, and a privilege.

"My first job with the company was sweeping floors on weekends. I was about 10 or 12. Mowed yards and helped with special deliveries. I'd go with Dad in the evenings. Sometimes hand him tools. Dad was handy with tools. Mike is too.

"In my teens I worked in the Wells' Sioux City office on Steuben Street pushing a broom, cleaning up after drivers, mowing the yard… any job the general manager, Marlin Idso, would have at the warehouse. (Idso now works in Le Mars.)

"Fred was open to letting us do what we wanted to do. If we didn't want to get into the business, he said it was okay."

Greg studied at Sunnyside Grade School and Hayworth Junior High in Sioux City and was graduated from East High School. Then on to Morningside College in Sioux City, finishing at University of Nebraska, Lincoln, in business administration.

In 1990 he joined Wells' as sales representative in Northwest Iowa… Orange City, Sioux Center, Algona, Spirit Lake, Rock Rapids, Spencer, Rock Valley. "That gave me a lot of knowledge about our products. I became familiar with the aspects of different products. Good proving ground."

In his next step, Greg went into merchandising as manager, with more responsibility. "Spent time with sales representatives from Iowa to Texas… giving them tools they needed to be better salespeople. In each region I worked with local merchandising and sales representatives, managers, who'd go out and do what I did."

Youngest of the third generation in the business, Greg now is Director of Business Development Merchandising.

Weekends with his wife Pam Sailer Wells and stepson "Nick" (Nicholas) Cronin are spent on the Missouri River. Greg was born near the river. "We enjoy animals, have a miniature Schnauzer and a Newfoundland. We like walking, camping out, swimming, water skiing and boating. Swimming is a big interest." Greg participated in competitive swimming for 10 years. "Still do," he says.

Moving to Le Mars in 1991, he married Pam in the Spring of 1996. They built their home in Le Mars. "She is civic-minded. A wonderful wife," says her husband.

Life is good on the river and in The Wells Spring that, since its humble start in 1913, has flowed through rapids and quiet pools for nearly 87 years.

Greg Wells, *(right)* ***is proud of his family, Pam his wife, and his stepson Nicholas, "Nick"*** *(left)*

(Photo courtesy of Greg Wells family)

Postscript

It is not how much we do, but how much love we put in the doing

Mother Theresa

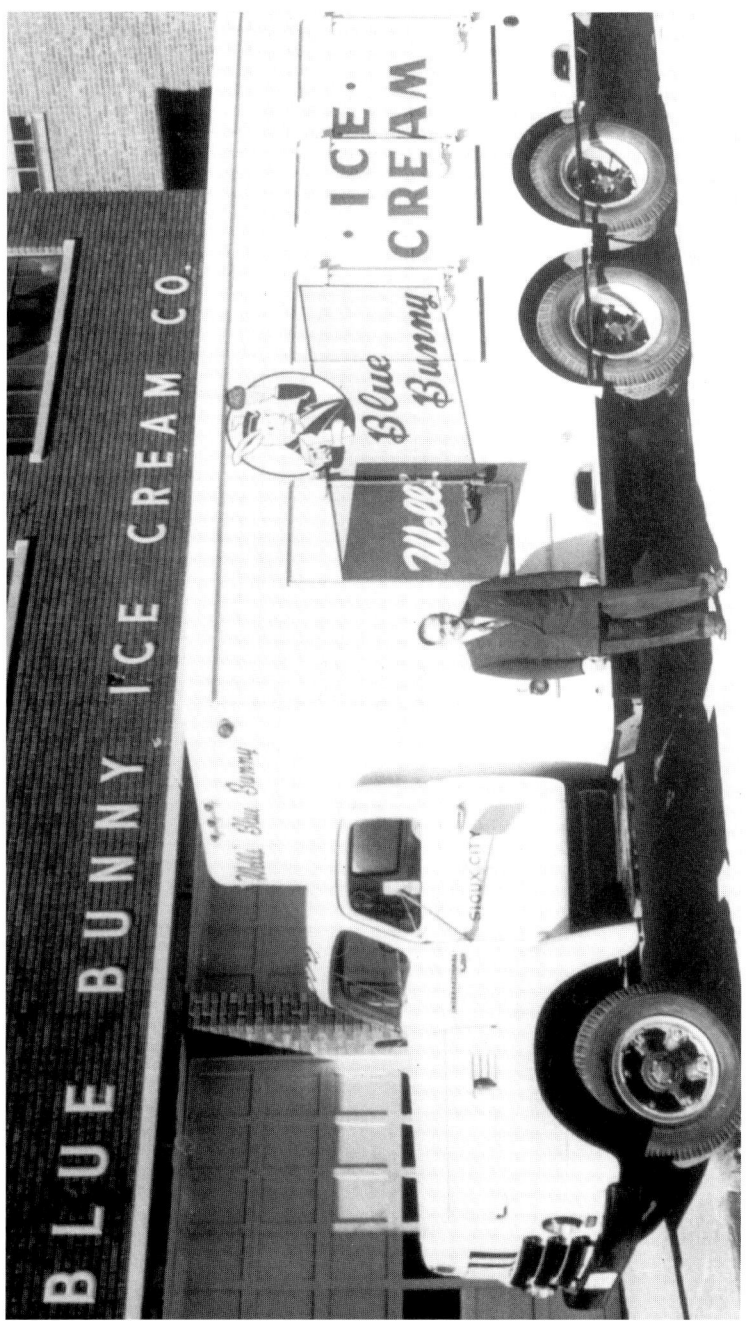

For more than seventy-five years the Wells' Blue Bunny name has remained the same. Its logos have changed in small ways, but the initial name persists. Here Roy Wells is shown with a truck in Sioux City

Long-Lived Logos

 Four men sat on comfortably overstuffed chairs at Sioux City's West Hotel on Nebraska and Third… scratching their heads. Before them a bushel basket brimmed with slips of paper and unopened envelopes. The fall day in 1935 was blustery and cold.

Wells' Dairy had run an ad in the Sioux City Journal that brought an unending response. A $25 prize was offered for a name sought by Wells'… a clever identification for its popular ice cream. Contenders sent suggestions that glutted the post office and burdened postmen.

Wells' officials were looking for a grabber, something that young and old would hook into… a name to be remembered.

On this chilly day, all morning long Fred Hooker Wells Jr.; his son Roy; cousin Harry C. Wells, head of the Sioux City outlet; and A. Bracklin, manager of Sioux City's Kresge store, sat opening mail and unfolding entries.

A pile designated "not suitable" grew. Another smaller one stacked up nearby.

"Superior," one of the men read from an entry. He tossed it on the larger pile.

"Super Quality," another read, adding it to the growing rejects. "Good but not good enough."

Noon neared. They had gathered early that morning. "How about this one," a third asked. "Wells' Blue Bunny Ice Cream?"

Four men stopped riffling through papers and paused to look up. Each sieved the idea and stirred it around in his thoughts.

"Pretty good," said the fourth. "Pretty good."

"We were looking for a company name," said Roy Wells more than 60 years later. That one caught their fancy. It stuck. It has been a winner.

Winner of the prize, George Vanden Brink, worked as an artist at the Sioux City Journal. He later became a publisher in Washington, D.C.

His wife, Lucille, has said it was their son, John who brought about the Blue Bunny designation. As a child John at Easter saw a Blue Bunny display in a Sioux City department store window. "John, delighted, said he wanted a blue bunny," triggering her husband's creative juices.

Later, Roy sent a miniature Wells' Dairy Blue Bunny truck to John. Now the little ancient truck sits on John's desk at his Des Plaines, Illinois office. "He's so proud of it," his mother says.

1935

The first Blue Bunny logo featured the ubiquitous bunny holding an ice cream cone and an early square carton of ice cream. He wore a suit, bow tie and short soda jerk hat between long ears. His eyes twinkled and his whiskers appeared newly trimmed. For thirty years that old bunny brought in business.

He was displaced in 1965 with a younger-looking bunny flashing a wide-open smile, outstretched arms ready to embrace the public, longer ears, longer whiskers and a short one-button vest.

1970

Only five years later our loving bunny with arms ready to hug was replaced. A new logo showed the Wells' Blue Bunny name more visibly inside a band of red, white and blue. Only bunny's friendly face, his big smile and long ears remained.

Again five years later in 1975 another bunny joined the logo bunnies. This one, dubbed "Ol' Blue," was designed for use on children's ice cream products. Also known as "True Blue," he had a toothy smile, a stylishly long, thin body, long legs, long arms. He looked a little like an Arabian shiek in a head-covering nearly lost between two enormously long ears. Ol' Blue stood tall, whimsically signaling to someone… probably kids.

After a long pause, in 1997, and after marketing, sales, and product planners studied long and hard, another logo

Long-Lived Logos

emerged, fit for the Twenty First Century.

Blue Bunny, now blue, rests his long, slim body on a bar of blue carrying the Blue Bunny legend and a round, red Wells' crest beneath. Gold bars give it a classy, modern look.

The phone rang some years ago in Roy Wells office. George Vanden Brink and his son were coming through Le Mars on their way "to the Lakes," Okoboji or Spirit Lake. They didn't say which.

"May we stop to see you," Vanden Brink asked.

"They didn't make it," Roy says. Travel problems developed.

*Vanden Brink missed a chance to hear how his bunny had developed and evolved during the 64 years since he conceived it.

But his wife has had a word to say: "Best ice cream there is!"

*The man who created the first Wells' Blue Bunny logo, used from 1926 to 1951, died October 17, 1998, at Westminster, Maryland. A native of Maurice, Iowa, he was 94.

◀ First bulk milk truck was used to pick up raw milk from farmers

▼ Biggest truck in the Wells' fleet at the time, (1950s) ran between Sioux City and Valentine, Nebraska

▲ Model A Ford ice cream truck was layered with a double wall of salt and ice for cooling. The logo was of gold-leaf edged in black. Sylvester Kale drove the truck

▶ Truck used to deliver milk to Westmar College

◀ Long-time Wells' employee, Ole Van Goor, began work at Wells' when he was 14 years old. Here his truck sported a gold-leaf logo. Van Goor worked for Wells' Dairy until his death March 10, 1981

Long-Lived Logos

◀ Ray Wells and Fred Wells Jr. with a truck used at the Spencer, Iowa plant, which was known as Wells' Sanitary Dairy

▶ This Wells' vehicle was the first electric refrigerated truck, vintage about 1946

▶ In Mobile, Alabama in July 1996, a driver poses beside a Well's Blue Bunny ice cream truck that traveled door-to-door

(Photo courtesy of Bill and B.J. Johnson)

What we learn with pleasure we never forget

Alfred Mercier

The Wells Spring

Ice Cream Capital

 Some would question and others would agree that Wells' Dairy, Inc. has made Le Mars, Iowa the Ice Cream Capital of the World®....No other firm produces as much ice cream in one location as Wells' Dairy. In Le Mars abides the top governing seat, the chief office of Wells' Dairy, Inc.... *the capital*.

Now consider that Wells' Dairy began with a mom and pop operation in 1913 and has grown, in 87 years, to provide ice cream and dairy products for 48 states and 30 countries throughout the world.

Equally amazing is that Wells' Dairy, Inc. is a family-owned industry in which 15 family members, all male (save one woman), have participated. Wells' Dairy is the largest remaining family-owned dairy in the U.S.A.

What is this ice cream product, so palatable, so tasty, that it has survived, along with the family and the business to become prestigious... producing 2,000 and more ice cream and milk-related packages and products distributed throughout the globe?

Are we asking that a family secret be unveiled?

What is no secret is how this may have begun.

Doris Wells Zimmerman tells a story that she heard from her Uncle Roy Wells, the family storyteller, the memory keeper:

"Aunt Miriam mixed up the ice cream mix and put it in the container and Freda Wells, (Harry's wife) and Ray Wells (Doris' father) froze it in the hand-cranked wooden tub freezer."

Family gatherings were topped off with ice cream. Children pumped away on Aunt Nellie's player piano, while Fred Jr. and Ray cranked away at the freezer.

"It makes sense to me," Doris says, "that Aunt Miriam would go to a cookbook for a recipe new in the 1920s."

The former Doris Wells poses a possibility that the recipe Aunt Miriam used may have come from Aunt Nellie Wells' *Household Searchlight Recipe Book* (published by Arthur Capper of Capper's Weekly.)

Here it is… for custard ice cream. The family would have used nothing but the best ingredients, the best recipe:

    2 *Eggs*
    2 *Cups Milk, Scalded*
    6 *Tablespoons Sugar*
    1 *Teaspoon Vanilla*
    ⅛ *Teaspoon Salt*
    1 ½ *Cups Evaporated Milk,*
         or Cream

Beat eggs until blended. Add sugar and salt. Mix well. Add milk slowly stirring constantly. Cook over hot water until mixture coats the spoon. Remove from fire at once. Chill. Add flavoring and milk or cream. Freeze. 8 servings.

The Wells Spring

Use rock salt in the freezer, *Household Book* recommended, and finely crushed ice. "8 parts ice to 1 part salt. Mechanical freezers," it was pointed out, "have trays in which desserts may be frozen."

A full page of directions told the cranker how to crank the freezer, describing the freezer as a "bucket or tub of wood or metal, to hold the ice and salt and a container of rust-proof metal for holding the ice cream mixture." In the container, it said, is "a dasher which is turned. This whips the mixture as it freezes and aids in preventing large crystals."

Fred Wells, who had only a third grade education, began to make ice cream for delivery in 1926. In its 1931 edition, *Household Searchlight* listed 21 differing ice cream recipes.

Among choices were brown bread (made with grapenuts,) junket (made with raspberry junket,) cornflake, ginger, black and orange (raisins and orange flavor,) prune, lemon fig, and of course, "plain ice cream."

This is telling us how popular ice cream had become.

It may explain Fred and Miriam Wells' inspiration for what became the Ice Cream Capital of the World.®

Ice Cream Capital

A March/April 1999 Wells' Dairy Rabbit Tracks publication cover featured this photo of Blue Bunny memorabilia. Note the glass milk bottles. The square milk carton (upper left) was introduced into the dairy industry by Wells'. In front are early ice cream scoops, and an ice cream sandwich maker

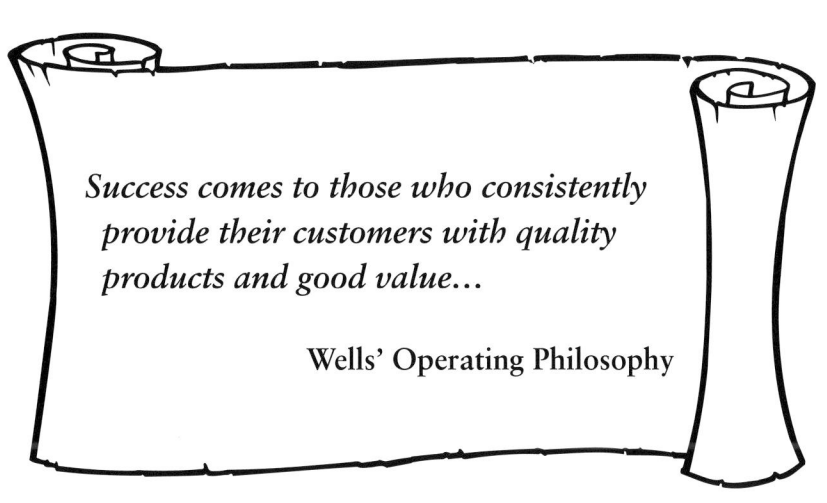

Success comes to those who consistently provide their customers with quality products and good value…

Wells' Operating Philosophy

When Fred Jr. and Miriam arrived in Le Mars in 1911 it looked prosperous compared to the drought-stricken area around Wellsburg from which they had come. The Opera House, a three-story building, is at center, left

(Photo courtesy Plymouth County Historical Museum)

In Summary

When Fred Hooker Wells, Jr. and Miriam, his wife, in 1911, rolled into Plymouth County, Iowa from South Dakota, wagon wheels grinding against the brick pavement in Le Mars, their team of horses lathered with sweat, he and Miriam, and his parents, wanted only to find a place to stop and rest.

Anxious, exhausted, downcast after 500 miles of dusty travel from Wellsburg, following wagon trails and railroad tracks, they arrived at a more prosperous area. This town looked good. The country and Le Mars appeared affluent compared to the barren life they lived in South Dakota.

The four had seen an automobile after they crossed the Big Sioux River at Hawarden. A Ford! A Model T Ford! The driver wore a helmet and goggles to keep out the dust. His companion clutched a bonnet and shawl.

No one they knew could afford the new Ford that was selling for $850.00. Dry years and a financial panic in 1907 had made such a purchase out of the question. But things might change for them now they were headed back to Chicago.

Some things did indeed change. For example, two years later Ford's moving assembly line brought prices for a new Ford down to $260.00. (A Ford Runabout with isinglass windows, a little later, carried a $475.00 tag.)

Le Mars looked good. Talking with a friendly overalled farmer downtown, Fred commented on the brick pavement.

"Ya," said the kindly soul, "first brick paving in the State of Iowa. Put it in in 1905." Fred was impressed. This must be a progressive area.

"There's talk of rural telephone lines, too," another resident told them. "See that bank there… one of the finest in Iowa, they say. German American Savings. Bedford stone."

Three banks here, Fred noted. That's a good sign.

The town had a college, Western Union; a county poor farm; a post office; a big hotel, The Union; a depot; two newspapers; a wheat mill; a redstone County Courthouse; Spotts and Post Drug Store; Dalton Opera House. The Marble Works, they learned, had just bought out three other marble plants. A country club had been built.

This was better than the Wells had seen in five or six years. Better than the towns they had come through in Dakota.

They decided to stay.

Good department stores in Le Mars pleased Miriam. Even so, the Montgomery Ward catalog still played an important part in the average life. People depended on the catalog for some purchases.

A story goes that one Midwestern farmer asked Sears and Roebuck's mail order house to send him "one wife, Model

12-12 on page 112. As soon as possible."

A woman in a remote area is said to have sent in an order for embalming fluid listed in the "Monkey Ward" catalog. But, she asked, "Must I pour it down my husband's throat just before he dies, or rub it in after he is dead. Please rush."

Wages averaged $2.40 a day. Child labor laws were adopted prohibiting children from working long hours or at night. A Federal Income Tax Law had been passed in Washington D.C., invoked by the Sixteenth Amendment.

As President, William Howard Taft distinguished himself when his 300-pound body got stuck in a White House bathtub.

The issue of Suffrage for Women heated up after Ex-President Teddy Roosevelt came out in its favor. Frightened males talked about "Petticoat Rule." In 1916, however, only 116 men cast ballots in Plymouth County, population 18,522.

Many in William Close's English Colony before 1900 brought prosperity from England to the area. Some moved farther West, or went back to England, dropping the number of Le Mars residents to a low of 4146. Now 15 years later population stood at 4500. The community was growing.

A preponderance of German-born, 2241, was changing the face of Plymouth County. Dutch and Irish immigrants followed closely in numbers.

These were a literate people. Of the more than 18,000 in the county, only 95 persons according to the record, were illiterate.

Attending school were 702 young persons. High schools in Iowa now were free. Josephine Winslow was principal of Le Mars

In Summary

Clark Street School.

Forty three churches in Le Mars served churchgoers… Lutherans, Presbyterians, Catholics, Episcopalians, Baptists and Dunkers.

Farmers were growing corn, wheat, oats, barley and rye and owned more than $1 million worth of farm machinery, more than 13,000 milk cows and 60,000 cattle, (steers, bulls and calves) 20,000 horses and mules, 2,500 sheep and, of course, swine. (W. S. Freeman *History of Plymouth County*)

Three railroads came through Le Mars and the county: the Illinois Central, the "Omaha" and the Northwestern.

James E. Kelley was first postmaster in the "fine new Federal Building" in Le Mars, the county seat.

John Starzl had purchased *Der Harold*, a German weekly newspaper of which Roman Starzl, his son, later became owner and publisher. Roman called it the *Le Mars Globe-Post* after the *Globe* and *Post* merged.

By 1918 a Primary Road Law was passed by the legislature in Des Moines. The state began to "come out of the mud." Paving of state roads began in 1925.

By this time Fred Wells' Dairy business was doing well. Wells' began to make ice cream. A quart of milk cost five cents; a pound of butter, ten cents. Fred had begun to add other dairy products to his milk delivery service.

Millions of copies of the *McGuffy Eclectic Reader* were sold during the 1920s. McGuffy taught that the primary purpose of government was protection of private property. (Wealth was then a sign of inner salvation; poverty was considered a sign of God's disapproval.) To succeed one needed to be "sober, frugal and energetic," according to the prevailing ethic.

Had Fred Jr. learned his philosophy of life from the *McGuffy Eclectic Reader*?

A handsome depot served several rail lines, commerce and travelers in the early 1900s

(Photo courtesy Plymouth County Historical Museum)

A new ice cream plant and high rise freezer, and an engineering plant now comprise part of 18 Wells' Dairy buildings in Le Mars

Soaring

> *One can never consent to creep*
> *when one feels the impulse to soar*
>
> Helen Keller

Although a man of vision, a good businessman, a progressive businessman, how could Fred Wells, Jr. have foreseen, how could he have looked 87 years down the road?

Only a crazy wild dreamer would have dared to project a maze of 18 ice-cream-white and bricked buildings, and milk silos rising up 60 feet into the air; housing ice cream loaded on pallets groaning with some 2000 packages at minus 20 degrees below zero temperatures; supervised by ice-cream-coated attendants and dark business-suited executives hovering over the whole like mother hens with new chicks.

Could that have been the way Fred Jr. planned it? Is that the way Fred Jr. envisioned it?

Let us try to see it as Fred Jr., in 1913, saw it... a way to feed and clothe his growing family, provide a roof to cover them and hope that by working hard, keeping a frugal stance, and offering the best that was in him for his customers and his family, he would have a better life for all.

It took a slow steady struggle just to survive. Records

In Summary

show that Fred Jr. paid off his initial investment of $250.00 in his mom-and-pop dairy business in $10.00 increments to Ray Bowers. Bowers supplied milk from 10 to 15 cows each day to meet Fred's delivery needs.

In three years, through toil and thrift, the business had gone well enough that a small concrete block building was erected at 121 Second Street near downtown Le Mars.

The family moved from a home at 1118 Franklin Street, near Plymouth County Courthouse and Jail to another home next door to the cement block structure. Four young sons began to get involved in the business. A closeness developed and an interest in the business grew as each son shouldered more responsibility.

In mid-1920, Fred's enterprising nature inspired him to start making ice cream. The business expanded as the product, made of the best quality ingredients, impeccable, appealed to the most fastidious tastes. Business flourished.

A few years later, Fred's brother, Harry C. Wells abandoned South Dakota to join Fred in a partnership venture at Sioux City. Then came an offer to sell to Fairmont Creamery. Wells' sold under a five-year contract. At expiration of the agreement Wells' Dairy returned to Sioux City. According to the contract Fairmont was prohibited from further use of the Wells' name.

In the meantime a contest produced the Wells' Blue Bunny logo that has been used with variation for 65 years. Growth continued to spurt.

Because of Fred Wells' frugality, Wells' Dairy weathered the Great Depression. Wells' had no indebtedness.

The 1940s was a time of transition. World War II claimed employees. After World War II, sons were moving into positions of greater responsibility and trust. More space was needed, more equipment. The concrete block building grew and grew. Seven additions eventually evolved.

In the 1950s Fred Wells, Jr. and his wife Miriam, his helpmate in the business and in life, died just two years apart. It was as though "a great tree had gone down."

Times were changing. Bells tolled again… as a poet once said. This time the need to pay cash for improvements at the dairy was passing. The new generation taking over looked at business with educated eyes.

During its 50th anniversary in 1963, milk products went out in 90 trucks to Nebraska, South Dakota, Minnesota and throughout Iowa. To process, sell and deliver, 150 employees were engaged. Wells' was in a fast-growth period.

By 1979 Wells' Dairy incorporated. A corporate office was born at 1 Blue Bunny Drive. Home for a greater fleet of trucks went up.

A milk plant in Omaha was dedicated in 1983. Fruit juices, yogurt and milk were processed in Omaha.

Two years later, a major expansion, the North Ice Cream and Milk Plant was undertaken. New production lines resulted. A central receiving warehouse followed and a computer controlled inventory system.

A Technical Center for Research and Development of new products was next.

"My father Fay," Dan says, "had the insight to foresee need for a research and development department. The low-fat,

In Summary

no-fat, no-sugar-added revolution was underway. Fay wanted to get ahead of the curve.

"We purchased a building and remodeled it. Eddie Oamen, Ph.D. was hired to run the new Research and Development Center.

"Two R&D projects were introduced: (1) Wells' Blue Bunny Lite 85® Yogurt, and (2) Aspartame Sweetened Yogurt. We took the market by storm.

"Further, Con Agra was contacted. Con Agra's Healthy Choice® low-fat ice cream was invented in Le Mars!"

Now a growing number of customers began coming to Wells' for the dairy's expertise. A contract packaging system, put Wells' products into packages that bore the names of giants such as Disneyland® and Healthy Choice®.

The corporate office, bursting at the seams, climbed upward another story.

"In the fall of 1991," Dan says, "I heard that Merritt Foods, a competitor, was closing. I made a phone call to inquire about purchasing the company, and was put in touch with the right people.

"Dave Wells and I attended an auction. There I met Doc Abernethy, who opened up a whole new world of opportunity for us.

"In 1955, Doc was co-inventor with originator James S. Merritt of Bomb Pop®. Doc also invented extruded novelties with character faces. He showed me all the extrusion nozzles invented over the years: one in particular… Teenage Mutant Ninja Turtle™! Doc led us into the street vending business through purchase of Merritt Foods.

"We paid $17,000 for the nozzles and opened a new division, vending. This, I felt, kept another competitor from entering the vending effort.

"Doc, Gary Wells and I often went to Los Angeles for signing with studios on the rights of characters we had purchased... to MGM, Larry Harmon and Surge Licensing.

"We went for the king of them all, Warner Brothers, for rights to Looney Tunes™ characters. We persisted. We got the agreement.

"Doc helped us form a relationship with Bob Holder who now works for Wells'. Together we put in place the Master Distributing agreements that form our marketing strategy. Ken Reuter, now Marketing Manager, Food Service Group, helped.

"Sadly," Dan says, "I attended the funeral of Doc Abernethy at Kansas City in 1994."

On Le Mars' south side 112 acres of farmland was acquired. An enormous freezing and manufacturing plant, distribution and truck center sprang up. More warehouses were needed, more production lines, more square footage... nearly a million and a half in time. Plans for future space were written into blueprints. Demand followed demand.

"They have come to us," Doug Wells has said. "We didn't have to go after them." A lifetime goal of providing customers with "quality products and good value," has paid off over the years.

Founder Fred's Puritan ethic of industry, sobriety and thrift, self-reliance and discipline could be questioned by no one.

In Summary

Dairy Foods Magazine has said of Wells' Dairy, Inc. that it has a "fantastic reputation."

Chairman and Chief Executive Officer Fay Wells notes that Wells' Dairy is the "largest remaining family-owned dairy in the United States." Among the first to innovate "lite, non-fat" products, Wells' also was first to co-pack. It is the largest co-packer in the nation.

Its recent contract with Haagen-Dazs/Pillsbury boosts that image. Frozen ice cream products are produced for Haagen-Dazs.

"Partly because of the new agreement," Doug Wells says, a 120,000 square feet expansion of South Ice Cream Plant's manufacturing facility was undertaken, completion October, 1999.

"Already the largest free-standing ice cream manufacturing plant in the world, the new expansion brings the facility to 550,000 square feet," according to Dan Wells.

Four hundred new jobs were created.

Gary Wells, Executive Vice President of Marketing: "We produce the most ice cream in one location in the world." Wells' is fifth largest in the U.S.A.

As Senior Vice President of Manufacturing, Doug Wells says, "Our computers know where every ice cream pallet is, what's on it, and when it was put there." Deeply involved as all Wells' family members are, he adds "It's cheaper to centralize and easier to maintain quality (by staying in Le Mars.)"

To that brother Dan Wells, Senior Vice President and Treasurer, Contract Packaging and Procurement comments, "I know about half of our employees by their first names. I grew up with most of them."

That is another factor in Wells' great success. Labor relations have always been good. Employees are made to feel part of the greater Wells' family. Employee suggestions are often adopted. Recognition is generous. Awards for accomplishment are presented. Special events are planned.

"Wells' pays special attention to all employees and takes a genuine interest in them," Wayne Kruse, Vice President of Cost Control has said. *Rabbit Tracks*, their newsletter, records weddings, births, deaths, and employee activities outside work. *The Inside Scoop* gives further details.

Six of Wells' employees have been with the business for 35 to 40 years. Forty seven have observed between 25 and 35 years with the company. Fifty four have been with Wells' from 20 to 25 years.

An example: Dan recalls "In 1995 I attended a retirement party for a fine employee, Earl McIntire. Earl began working for Wells' in 1970 after we purchased Lakeside Dairy Depo near Okoboji/Spirit Lake. I rode the lake regions door-to-door with Earl doing the book work for each stop. At the party I was proud I had helped teach him. And I suddenly realized I was no youngster anymore."

Corporate Wells' also is involved in community affairs. It has supported Little League, Plymouth County Fair, Plymouth County Historical Museum, DARE (drug abuse), made Wells Family Foundation scholarship awards, bought equipment for local parks, provided ice cream treats at band concerts and local festivities, trucked donations to Grand Forks, N.D. for flood relief and to Miami and the East Coast

In Summary

for hurricane relief. It served on Milestone Partners during Iowa Centennial programs.

In turn, 200 community volunteers and school children helped plant 2000 trees and shrubs in the South Ice Cream Plant area.

Finishing their morning coffee and closing their back doors in the year 1999, some of Wells' 2200 employees leave their homes, climb into their cars and daily head toward one of Wells' Dairy's 18 locations, plants, offices, and other buildings scattered throughout the city and other points around the country. Some of those 2200 arrive home from shift work at the same time.

At some locations arms are thrust into white gowns, caps or hair nets donned, and feet pushed into boots… work that insures purity of milk from 70,000 cows that comes pouring in daily to be processed.

Some pull up in front of the Research Center, the Engineering Center, the South Ice Cream Plant, the North Ice Cream Plant, warehouses, truck bays.

Many may mount one of some 200 semi-trailer trucks loaded to the hatches with more than 2000 ice cream products and packages of all sizes and brands destined for wholesalers and retailers across country.

Behind the wheel they roar off out of town, one truck every ten minutes. Waiting are customers from Stevens Point, Wisconsin to Seattle, Washington; from Sumatra to Florida… waiting for Cyberbyte®, Bomb Pops®, Pink Panther™, ice cream sandwiches, push ups, ice treats, milk, sour cream, cottage cheese, Mickey Mouse bars, Mississippi Mud, drumsticks, Health Smart®, The Champ!®, more than a billion and a quarter items.

Some are bound for the Pacific Coast, for Japan or Brunei, Malaysia, Singapore, Saudi Arabia, North Dakota, Mexico, Russia.

All the while these products are freezing comfortably behind the driver in an enormous cab bearing the Wells' Blue Bunny logo. The World Wide Web at the same time provides food industries with information about Wells' Dairy.

From Haagen-Dazs to General Mills to Weight Watchers to Con-Agra all are waiting daily for products that are known as reliable in taste, appearance and quality.

They don't wait for long. Customer service and value for your $$$ has always been a watchword for Wells' Dairy, Inc. ever since Fred H. Wells Jr. left his home before 8 a.m., alone in 1913, with only an old gray horse and a delivery wagon to take milk door-to-door.

In 1997 this mega business in Le Mars produced 55 million gallons of ice cream and frozen desserts and 100 million dozen novelties and was named Ice Cream Capital of the World by the State of Iowa. That, obviously, is not just rhetoric.

In plans for the future, there is talk of using robots for milk and ice cream plant operations. Truth is, some use of robot equipment already has begun.

If Fred Jr. could only see this! Would he scratch his head and ask "Do we have the money to buy these crazy things."

In Summary

By 1938 the original cement block building uptown had been enlarged and converted to a more modern facility

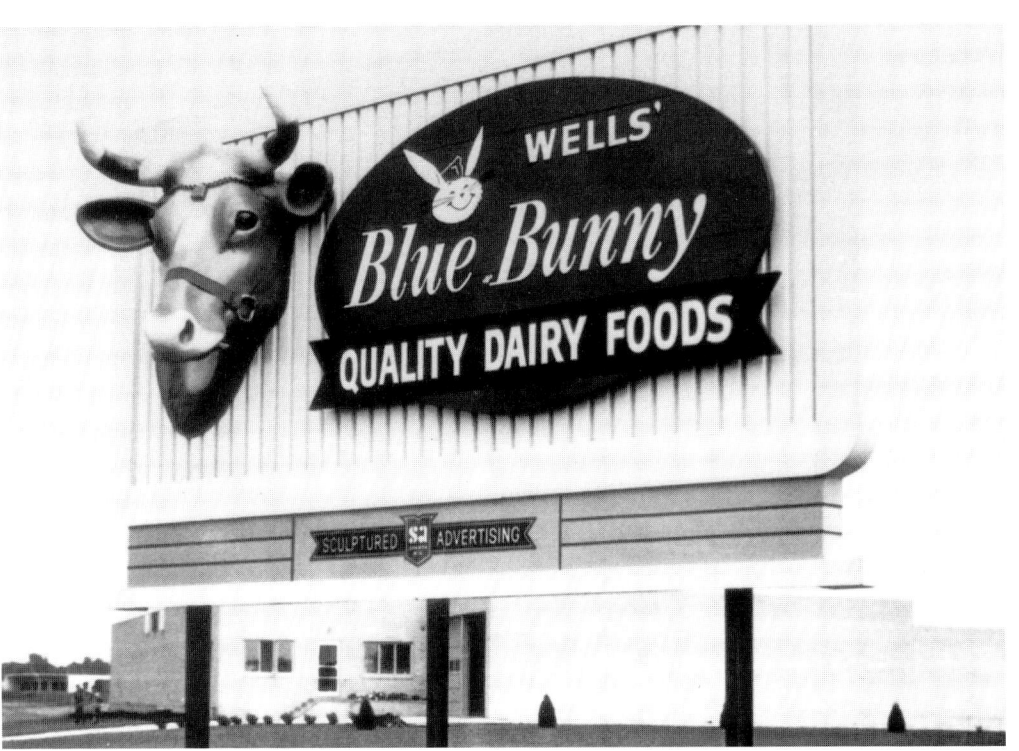

In 1963 a new milk plant opened at Twelfth and Blue Bunny Drive. The attention-grabbing cow's head on a billboard above the building grabbed the attention of Westmar College students who stole it one dark night, as Roy Wells tells the story

Rabbit Tracks, the company newsletter was first issued 20 years ago and has grown in coverage and interest

In Summary

Wells' Dairy Inc. donated a generator and engine to Sioux City's Lutheran Hospital in 1963

In 1962 Roy Wells (left) and Fred D. Wells, were snapped at Le Mars Airport with a Wells' Dairy plane

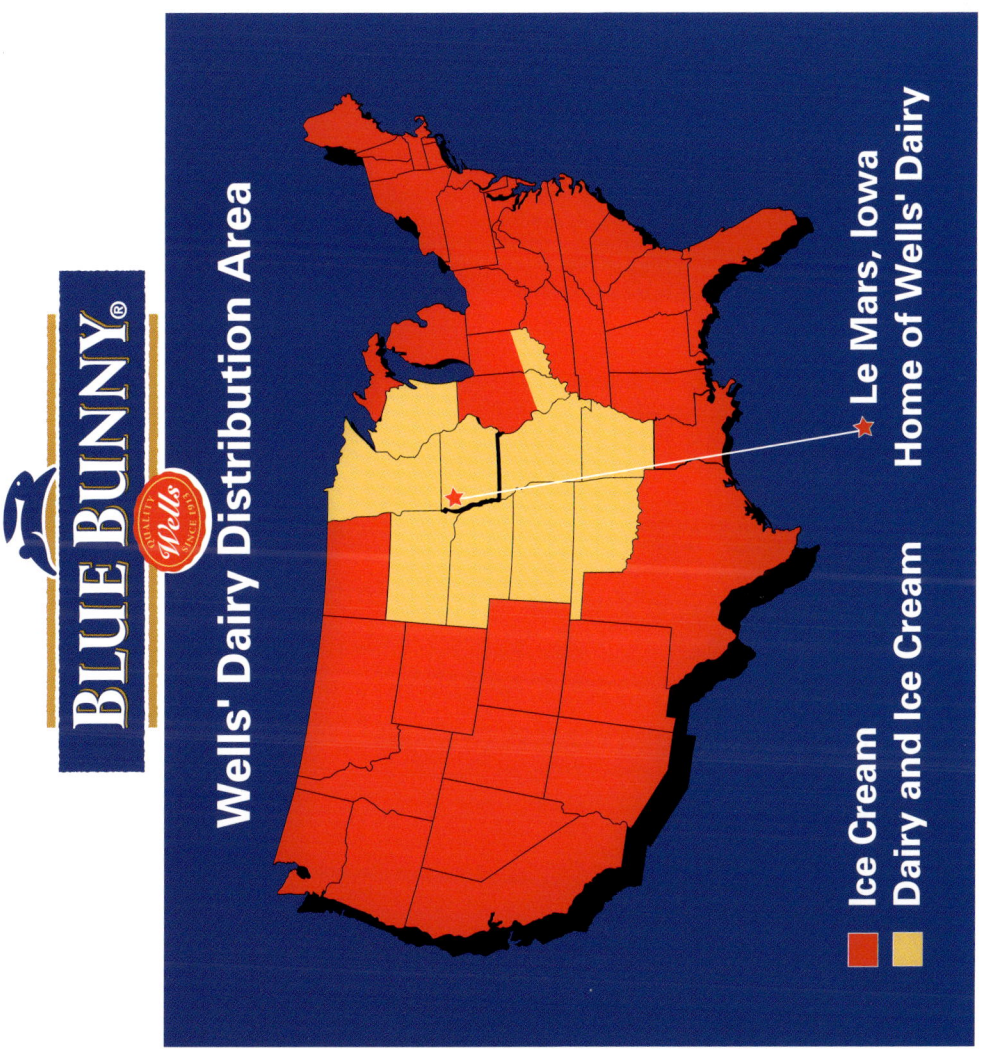

Phenomenal growth of Wells' Dairy by 1998 found sales offices in 48 states, and 30 countries world-wide. The milk from 80,000 cows each day is required. 2200 Wells' employees make distribution of its products possible

In Summary

A former Le Mars resident, Bill Johnson, purchased Wells' Blue Bunny products on an Alabama trip in 1996 and forwarded the evidence

On the 50th anniversary of the ice cream sandwich, June 1995, Dan, Gary and Doug Wells were in Washington, D. C. to create the World's Largest Ice Cream Sandwich

The Inside Scoop

Dimensions:	8ft long x 3ft wide x 1ft high
Total weight in pounds:	830 lbs
Total pounds of ice cream used:	780 lbs
Total gallons of ice cream used:	131
Total pounds of dough per wafer:	57 lbs
Flour	120 Cups
Sugar	36 Cups
Vegetable Oil	10 Cups
Cocoa	1 Cup
# of regular ice cream sandwiches that could fit inside World's Largest:	4,320

— An American Classic —

In the Wells' Blue Bunny corporate family today are (front, left) Gary Wells, Fred D. Wells, Fay Wells, and (second row) Dan Wells, Dave Wells, Doug Wells, Greg Wells and Mike Wells

(Photo courtesy of Wells' Dairy)

In Summary

Doc Abernethy, Dan Wells, Bugs Bunny, Nick DeRose and Tweety Bird are shown in front of a Wells' presentation at a National Vending Convention in 1992

(Photo courtesy of Dan Wells)

Fay Wells, right is shown with Dr. Ed Oamen, left, and Dr. Verner Nielsen, head of dairy science at Iowa State University. Dr. Oamen, Director of Research and Development has been with Wells' for more than 10 years

(Photo courtesy of Doug Wells)

Wells' Dairy has expanded into the Russian market. Wells' produces for 30 countries world-wide, including Japan, Saudi Arabia, Singapore, Puerto Rico, Nigeria, India

(Photo courtesy of *Le Mars Daily Sentinel*)

In Summary

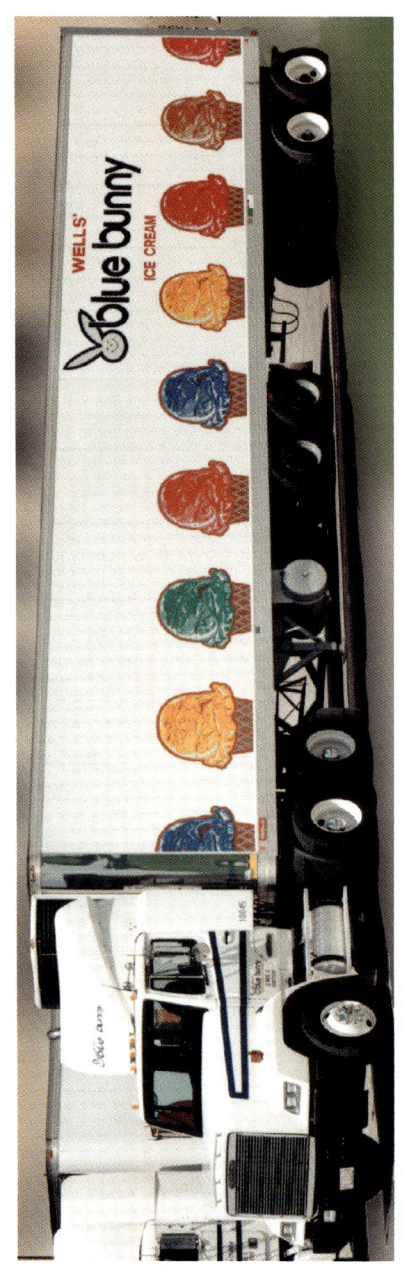

Time Magazine in 1992 ran a story that included this Wells' truck in the background. Dan Wells, was shown in the foreground, and was described as "Wells' president," the designation referred to in the letter from Ann McAnally. (opposite page)

June 30, 1997

Dear Mr. Dan Wells,

    I seem to remember Time Magazine designated Dan Wells as president.

    Seeing your commercials several times a day on national television has really taken me back in time. From 1938 to 1942 I was the "Blue Bunny Girl" in Sioux City, acting as receptionist, telephone person and accounts receivable bookkeeper. Elmer Rustwick functioned as Office Manager, and Harry Wells had a cubicle of his own, where he conducted business in a cloud of cigar smoke. During the summer, another girl was added, to help with the seasonal rush. One of these young ladies, come September, married one of the drivers!

    The crew in the backroom was composed of Ray Kirby, the Osepowicz brothers - Eddie and Johnny- Mid Blair (he married the Summer Girl,) and one other gentleman whose name escapes me, Plus an occasionally-sober "Emil", the refrigeration repair man. These lads worked twelve-to-fourteen hour days, six or even seven days a week.

    My starting salary was $55 a month. In four years this had increased to $85. Harry offered me the munificent sum of $100 if only I'd forget the foolish idea of leaving to get married. Luckily it proved to be a good marriage; we'll soon celebrate our 55th Anniversary and I've never had cause to regret my decision.

    When I started work, I believe the plant was on lower Pierce. One day Harry called me into his office and excitedly showed me a check for $10,000, his down payment on "our own building"! He said I'd probably never again see that much money all at once. Our new home was on Jackson, just below the Badgerow Building.

    We frequently had visits from Fred Wells or his son Roy, down from LeMars. And one day I was introduced to a man from Omaha who was soliciting outlets for his new line of frezen foods; Clarence Birdseye.

    I note with interest that you offer a new flavor each week. I remember Harry agonizing for weeks about the addition of one flavor - Chocolate Revel.

    I have seen a newspaper article that referred to your "Board of Directors" even a Time Magazine article proclaiming your status as the leading independently owned Ice Cream Company. You have evidently done fine without me; and I can only wish you continued prosperity.

The "Original Blue Bunny Girl"

Ann McAnally

Ann (Knudsen) McAnally

                            (By permission of Ann McAnally)

In Summary

This five-axle tanker bearing the newest Blue Bunny logo has pulled up with a load of raw milk in front of Wells' milk silos

(Photo by Joanne Glamm, courtesy of *Le Mars Daily Sentinel*)

In Summary

Memory Notes

About the Author

An Iowa native, Fran Sandrock (nee Frances Johnson) launched her writing career after graduation from Le Mars High School by writing policy amendments for Le Mars Mutual Insurance Company.

Following marriage she pursued a career in journalism working for weekly newspapers in Ohio. Moving through all the editorial jobs on a newspaper, she became managing editor for one of the largest weekly papers in Ohio, part of The News Sun chain in the Cleveland area. She also published in Iowa and Ohio magazines.

Among many professional awards made to Sandrock were: National Federation of Press Women (NFPW) Top Ten Winner, Southern Illinois University Golden Dozen Award, Ohio Buckeye Press and Arkansas Writer's Conference awards.

She is past president of Ohio Press Women and served on the NFPW board. While still in Iowa she was elected president of Sioux City Business and Professional Women.

For more than nine years she has written a weekly "Recollections" column for *Le Mars Daily Sentinel* covering historical musings of Northwest Iowa and family anecdotes about her early life as part of a large extended Iowa family.

Sandrock studied at University of Akron and Wayne General and Technical College in Ohio, and Saddleback College in California (advanced creative writing).

The mother of two daughters and a son, she has six grandchildren and a great-grandchild. She lives in Leisure World, Laguna Woods, California.